COMMUNION

'When you read this incredible story, do not be too sceptical: somewhere in your own past there may be some lost hour or strange recollection that means that you also have had this experience.'

COMMUNION is one man's powerful testimony of his shattering encounters with intelligent non-human beings who invaded his home and left him doubting his own sanity. Whether you believe his story or not, you will be compelled to read every word. Who are these visitors? Where do they come from? Why are they here, and why do they stare at him, seeking the depths of his soul—seeking communion?

'This book is about forming a new relationship with the unknown. Instead of shunning the darkness, we can face straight into it with an open mind. When we do that, the unknown changes. Fearful things become understandable and a truth is suggested: the enigmatic presence of the human mind winks back from the dark.'

Whitley Strieber

COMMUNION

A True Story
Encounters with the Unknown

ARROW BOOKS

Arrow Books Limited
62-65 Chandos Place, London WC2N 4NW

An imprint of Century Hutchinson Limited

London Melbourne Sydney Auckland
Johannesburg and agencies throughout
the world

First published in Great Britain by Century 1987
Arrow edition 1988
Reprinted 1988

Printed and bound in Great Britain by
Anchor Brendon Limited, Tiptree, Essex

ISBN 0 09 953420 7

To the ones who have slipped into the mirror,
And the ones who reflect it in their eyes.
To the ones who must hide everything,
And the ones who lose what they hide
To the ones who cannot be silent,
And the ones who must lie.

Acknowledgments

I have been privileged to have the help of many accomplished members of the scientific community. I wish to thank Donald F. Klein, MD, director of research at the New York State Psychiatric Institute and professor of psychiatry at the College of Physicians and Surgeons, Columbia University, for his expert hypnosis. Dr. Robert Naiman was kind enough to provide similar support for my wife. Dr. John Gliedman offered skillful scientific evaluation of my ideas, and his own essential contributions. Dr. David Webb, recently a member of the National Commission on Space and presently professor and chairman of space studies at the Center for Aerospace Sciences at the University of North Dakota, and Dr. Brian O'Leary, astronaut and planetologist, provided insights that combined expert knowledge, healthy skepticism, and a firm adherence to the known, without which I could never have completed my task. Dr. Bruce Maccabee, research physicist with the United States Navy, read my manuscript for its physics content, but any errors in this area remain my own. David M. Jacobs, Ph.D., associate professor of history at Temple University, was kind enough to offer his comments as well, with special reference to historical background.

I would especially like to thank Budd Hopkins, who has sacrificed an enormous amount of time and effort—often approaching the heroic—for myself and those like me who have been dragged to the edge of reality.

The assistance of the physicians and scientists who have advised me does not imply that they support my conclusions, such as they are, about what happened to me. Their interest arises from a desire to study what appears to be an unknown or misunderstood phenomenon. To the scientific community, the nature of this phenomenon remains an unresolved question.

Contents

> *The concrete world has slipped through the meshes of the scientific net.*
> —ALFRED NORTH WHITEHEAD,
> *Modes of Thought*

PRELUDE

The Truth Behind the Curtain

This is the story of one man's attempt to deal with a shattering assault from the unknown. It is a true story, as true as I know how to describe it.

To all appearances I have had an elaborate personal encounter with intelligent nonhuman beings. But who could they be, and where have they come from? Are unidentified flying objects real? Are there goblins or demons . . . or visitors?

At first, I thought I was losing my mind. But I was interviewed by three psychologists and three psychiatrists, given a battery of psychological tests and a neurological examination, and found to fall within the normal range in all respects. I was also given a polygraph by an operator with thirty years' experience and I passed without qualification. I had been indifferent to the whole issue of unidentified flying objects and extraterrestrials; I had viewed them as a false unknown, easily explainable as misperceptions or hallucinations. Now what was I to think?

The visitors marched right into the middle of the life of an indifferent skeptic without a moment's hesitation.

Later I found a large number of people who have had experiences similar to mine. Most of them were mentally stable. They did not cluster in any particular population group, but formed a cross section of American society. I have met, among many others, a scientist, a policeman, and a federal officer who have had encounters.

In my case there were witnesses, and physical aftereffects that are hard to ignore. Either what is happening is that visitors are actually here, or the human mind is creating something that, incredibly, is close to a physical reality. Whatever it is, it is not presently understood by science.

I know how it feels and looks to be with these visitors. I know how they sound when they talk and what it looks and smells like in their places. I know how they act and how they appear. I may even know something about why they are here and what they want from us.

Seeming encounters with nonhuman beings are not new; they have a history dating back thousands of years. What is new, in this latter part of the twentieth century, is that the encounters have taken on an intensity never before experienced by humankind.

What happened to me was terrifying. It seemed completely real. It was in clear, normal memory. Most of it was already present in my mind before I was hypnotized to aid recall.

There has been a lot of scoffing directed at people who have been taken by the visitors. It has been falsely claimed that their memories are a side effect of hypnosis. This is not true. Most of them started with memories and undertook hypnosis to attempt greater recall.

Scoffing at them is as ugly as laughing at rape victims. We do not know what is happening to these people, but whatever it is, it causes them to react as if they have suffered a great personal trauma. And society turns away, led by vociferous professional debunkers whose secret fears apparently close their minds. Other, more responsible scientists are very legitimately concerned that serious pursuit of an answer to the enigma of unidentified aerial objects and possible visitors may lead science into study of a false unknown.

At least from a behavioral point of view, however, this can no longer be considered a false unknown. Something is happening, and intellectually well-grounded people need not shun it. Instead, the unknown can be faced with clear and open curiosity. When this is done something strange happens: The unknown changes. The enigmatic presence of the human mind winks back from the dark, and a little progress toward real understanding is made.

I suffered with this experience. Others suffered, and are still suffering. It is essential that effective support be developed to aid those who have it. The scoffing has to stop. I am ashamed to say that I have done it myself, in the past, at least vicariously. As far as "UFOs" went, I was with the skeptics.

I look up toward the night sky, visible through two high arches above my office windows. Almost all the way to the top of the arches the clouds glow with Manhattan's light. At the pinnacle there is darkness, and it draws me. I'm not only scared and upset, frankly I'm also curious. I want to know what's going on out there. As I watch, the night sky grows a little darker.

People who face the visitors report fierce little figures with eyes that seem to stare into the deepest core of being. And those eyes are asking for something, perhaps even demanding it.

Whatever it is, it is more than simple information. The goal does not seem to be the sort of clear and open exchange that we might expect. Whatever may be surfacing, it wants far more than that. It seems to me that it seeks the very depth of the soul; it seeks communion.

ONE

> When I had journeyed half of our life's way,
> I found myself within a shadowed forest,
> for I had lost the path that does not stray.
>   Ah, it is hard to speak of what it was,
> that savage forest, dense and difficult,
> which even in recall renews my fear:
>   so bitter—death is hardly more severe!
> But to retell the good discovered there,
> I'll also tell of other things I saw.
>
> —DANTE, *Inferno*, Canto I

THE INVISIBLE FOREST

First Memories

December 26, 1985

My wife and I own a log cabin in a secluded corner of up-state New York. It is in this cabin that our primary experiences have taken place. I will deal first with what I remember of December 26, 1985, and then with what was subsequently jogged into memory concerning October 4, 1985. Until I sought help, I remembered only that there was a strange disturbance on October 4. An interviewer asked if I could recall any other unusual experiences in my past. The night of October 4 had also been one of turmoil, but it took discussions with the other people who had been in the cabin at the time to help me reconstruct it.

This part of my narrative, covering December 26, is derived from journal material I had written before undergoing any hypnosis or even discussing my situation with anybody.

When I was alone, this is what it was like.

Our cabin is very hidden and quiet, part of a small group of cabins scattered across an area served by a private dirt road, which itself branches off a little-used country road that leads to an old town that isn't even mentioned on many maps. We spend more than half of our time at the cabin, because I do most of my work there. We also have an apartment in New York City.

Ours is a very sedate life. We don't go out much, we rarely drink more than wine, and neither of us has ever used drugs. From 1977 until 1983 I wrote imaginative thrillers, but in recent years I had been concentrating on much more serious fiction about peace and the environment, books that were firmly grounded in fact. Thus, at this time in my life, I wasn't even working on horror stories, and at no time had I ever been in danger of being deluded by them.

We were having a lovely Christmas at the cabin in late December 1985. On Christmas Eve there was snow, which continued for two more days. My son had discovered to his delight that the snow would fall in perfect crystalline flakes on his gloves if he stood still with his hands out.

On December 26 we spent a happy morning breaking in his new sled, then went cross-country skiing in the afternoon. For supper we had leftover Christmas goose, cranberry sauce, and cold sweet potatoes. We drank seltzer with fresh lime in it. After our son went to bed, Anne and I sat quietly together listening to some music and reading.

At about eight-thirty I turned on the burglar alarm, which covers every accessible window and all the doors. For no reason then apparent, I had developed an unusual habit the previous fall. As secretly as ever I made a tour of the house, peering in closets and even looking under the guest-room bed for hidden intruders. I did this immediately after setting the alarm. By ten o'clock we were in bed, and by eleven both of us were asleep.

The night of the twenty-sixth was cold and cloudy. There were perhaps eight inches of snow on the ground, and it was still falling lightly.

I do not recall any dreams or disturbances at all. There was apparently a large unknown object seen in the immediate vicinity at approximately this time of month, but a report of it would not be published for another week. Even when I read that report, though, I did not relate it in any way to my experience. Why should I? The report attributed the sightings to a practical joke. Only much later, when I researched it myself, did I discover how inaccurate that report was.

I have never seen an unidentified flying object. I thought that the whole subject had been explained by science. It took me a couple of months to establish the connection between what had happened to me and possible nonhuman visitors, so unlikely did such a connection seem.

In the middle of the night of December 26—I do not know the exact time—I abruptly found myself awake. And

I knew why: I heard a peculiar whooshing, swirling noise coming from the living room downstairs. This was no random creak, no settling of the house, but a sound as if a large number of people were moving rapidly around in the room.

I listened carefully. The noise just didn't make sense. I sat up in bed, shocked and very curious. There was an edge of fear. The night was dead still, windless. My eyes went straight to the burglar-alarm panel beside the bed. The system was armed and working perfectly. Not a covered window or door was opened, and nobody had entered—at least according to the row of glowing lights.

What I did next may seem peculiar. I settled back in bed. For some reason the extreme strangeness of what I was hearing did not rouse me to action. Over the course of this narrative this sort of inappropriate response will be repeated many times. If something is strange enough, the reaction is very different from what one would think. The mind seems to tune it out as if by some sort of instinct.

No sooner had I settled back than I noticed that one of the double doors leading into our bedroom was moving closed. As they close outward, this meant that the opening was getting smaller, concealing whatever was behind that door. I sat up again. My mind was sharp. I was not asleep, nor in a hypnopompic state between sleep and waking. I wish to be clear that I felt, at that moment, wide awake and in full possession of all my faculties. I could easily have gotten up and read a book or listened to the radio or gone for a midnight walk in the snow.

I could not imagine what could be going on, and I got very uneasy. My heart started beating harder. I wasn't settled back anymore; I was sitting up, a question just forming in my mind. What could be moving the door?

Then I saw edging around it a compact figure. It was so distinct and yet so completely, impossibly astonishing that at first I could not understand it at all. I simply sat there staring, too stunned to move.

Months and months later, I discovered that another person who has had the visitor experience first encountered it through the medium of this same peculiar figure rushing toward her in exactly the way that this one now rushed toward me.

Before I narrate those next few seconds, though, I would like to give an exact description of how the figure looked to me. First, I will describe the physical conditions under which I was seeing it. The room was dim but not dark. The burglar-alarm panel alone emitted enough light for me to see. In addition, there was snow on the ground and that added some ambient light. Had it been a person peeking into the room, I could have made out his or her features clearly.

This figure was too small to be a person, unless a child. I have measured the approximate distance that the top of the head was from the ground, based on my memory of the figure's position in the doorway, and I believe that it was roughly three and a half feet tall. altogether smaller and lighter than my son.

I could see perhaps a third of the figure, the part that was bending around the door so that it could see me. It had a smooth, rounded hat on, with an odd, sharp rim that jutted out easily four inches on the side I could see. Below this was a vague area. I could not see the face, or perhaps I would not see it. A few moments later, when it was close to the bed, I saw two dark holes for eyes and a black down-turning line of a mouth that later became an O.

From shoulder to midriff was the visible third of a square plate etched with concentric circles. This plate stretched from just below the chin to the waist area. At the time I thought it looked like some sort of breastplate, or even an armored vest. Beneath it was a rectangular appliance of the same type, which covered the lower waist to just above the knees. The angle at which the individual was leaning was such that the lower legs were hidden behind the door.

I was quite shocked, but what I was seeing was so strange I had to assume that it was a dream. Maybe this is why I continued to sit in bed, taking no action. Or perhaps my mind was already under some sort of control.

In any case, I sat there frightened but unable or unwilling to deal with what I was observing. My mind explained my vision to me: Despite my full wakefulness, it must be a hypnopompic hallucination. Such phenomena sometimes occur as one drifts between waking and sleep. I assumed that some minor disturbance had awakened me and I was experiencing such an illusion, and never mind the fact that I felt fully awake.

Because of its isolation, the house not only had a burglar alarm but contained a shotgun, which was not far from the bed at the time. Was that why the thing behind the door was wearing a shield, if that was indeed what it was? I have subsequently wondered if an earlier reconnaissance of the house might not have taken place and revealed the presence of the weapon?

The previous July we'd had an experience that should be reported here. I was reading at about half past eleven at night, when I distinctly heard footsteps—normal, human-sounding footsteps—move stealthily down our front porch to the area where I had just had a motion-sensitive light installed. The peculiar thing about these footsteps was that they came from the pool area and moved toward the road, the opposite of the direction that they would have come if it was a prowler from the road. At the time, I thought to myself that I would take the gun and go downstairs if the light came on.

No sooner had I thought that than it did. I dashed downstairs but saw nobody even though the light was still on. As it was attached to a fifteen-second timer, I found this startling. I had gotten out onto the porch in no more than ten seconds, and there was no place for an intruder to hide between the house and the road, not in that short time.

A careful investigation, shotgun in hand, uncovered nothing. I had been certain that I would see whomever it was running off. At the time I even entertained the notion that they must have jumped onto the roof, but there was nobody there.

Subsequently the light never worked right, although it

was in good order earlier that very evening. In September I took the bulbs out. Later in the fall the unit was replaced.

The next thing I knew, the figure came rushing into the room. I recall only blackness after that, for an unknown period of time. I don't remember falling asleep or lying awake. What I do remember is far, far more disturbing. My next conscious recollection is of being in motion. I was naked, with my arms and legs extended, as if I had been frozen in mid-leap. I was moving out of the room. There was no physical sensation at all, not of being touched, not of being warm or cold. I could feel myself as a shape and a mass, but not in terms of sensation. It was as if I had become profoundly paralyzed. Although I wanted desperately to move, I could not.

Because of my state of apparent paralysis, I am afraid that I cannot report that I was floating along on some magical pallet or a flying carpet. It could easily be that I was being carried. In any case, I was at this point in a state of panic. Gone was any fleeting thought of dream or hallucination. Something was hideously wrong, so wrong that my mind went blank. I couldn't think. Even if I had been able to make a sound, which I doubt, I couldn't try.

I must have blacked out again, because I have no further memories of being moved. The next thing I knew, I was sitting in a small sort of depression in the woods. It was quite dark, and frozen creeper was pressing tightly around me. I remember being startled that there was no snow on the gray earth.

I sat with my legs partly bent and my hands in my lap. Although I cannot recall this in any detail, I may have been leaning against something. I was still absent sensation. Across the depression to my left there was a small individual whom I could see only out of the corner of my eye. This person was wearing a gray-tan body suit and sitting on the ground with knees drawn up and hands clasped around them. There were two dark eyeholes and a round mouth hole. I had the impression of a face mask.

I felt that I was under the exact and detailed control of

whomever had me. I could not move my head, or my hands, or any part of my body save for my eyes. Despite this, I was not tied.

Immediately on my right was another figure, this one completely invisible except for an occasional flash of movement. This person was working busily at something that seemed to have to do with the right side of my head. It wore dark-blue coveralls and was extremely fast.

The depression appeared to be no more than four feet in diameter, but my eyes were not functioning normally— maybe for no other reason than that I wasn't wearing my glasses. (I am mildly nearsighted.) While the presence of others remains vague in my mind, the individual to my left made a clear impression. I do not know why, but I had the distinct feeling that this was a woman, and so I shall refer to her in the feminine.

She was as small as the others, and appeared almost bored or indifferent. I also felt that she was explaining something to me, but I cannot remember what it was.

I then saw branches moving past my face, then a sweep of treetops. I looked down, and below me the whole tall forest was corkscrewing slowly to the right. There was no chance to question how in the world I had gotten above the trees. I only saw and recorded. Then a gray floor obscured my vision, slipping below my feet like an iris closing.

The next thing I knew, I was sitting in a messy round room. My impression is that at this point I was actually being cradled by these people, as if they were aware of what was about to transpire. Movement to this totally unfamiliar environment, so suddenly and under these extremely unusual conditions, stripped away whatever reserves of collectedness I still possessed. While I had up until that point been able to retain a degree of control of my attention, this now left me and I became entirely given over to extreme dread. The fear was so powerful that it seemed to make my personality completely evaporate. This was not a theoretical or even a mental experience, but something profoundly physical.

"Whitley" ceased to exist. What was left was a body in a state of raw fear so great that it swept about me like a thick, suffocating curtain, turning paralysis into a condition that seemed close to death. I do not think that my ordinary humanity survived the transition to this little room. I died, and a wild animal appeared in my place. Not everything was gone, though. What remained, although small, nevertheless was occupied with an essential task of verification. I was looking around as best I could, recording what I saw.

The small, circular chamber had a domed, grayish-tan ceiling with ribs appearing at intervals of about a foot. I had an impression that it was messy, a living space. Across the room to my right some clothing was thrown on the floor. As a matter of fact, the thought even crossed my mind that the place was actually dirty. It was close and confining for me. The whole scale of it was small, tight, and enclosed. I seem to remember that the room was stuffy and the air quite dry, so it could be that the numbness of panic was wearing off.

Tiny people were now moving around me at great speed. Their quickness was disturbing, and in a curious way ugly. I had the thought that I was being taken away, and remembered my family. An acute, gnawing feeling of being in a trap overcame me. It was a truly awful sensation, accompanied as it was by the sense that I was absolutely helpless in the hands of these strange creatures.

Despite my extreme terror, I was aware of my surroundings. I know that I was seated on a bench, leaning against a wall. The predominant colors were tan and gray. The bench was the same color as the walls, and was rimmed by a lip of dark brown. From the clarity of my memory of these rather muted colors, I surmise that the room was lit, although I did not see the source of the light.

There was something quite beautiful, I think, having to do with a lens in the ceiling, but I can remember little about it. Perhaps there was a lens at the point of the ceiling, through which some colorful scene could be observed.

There is no way to be certain of how long I remained in

this room. It seemed to be a stay of no more than a few minutes or even seconds. It may have been longer, though, because I had time to look around me and note numerous details. While I had before been totally paralyzed, I was now able to move at least my eyes and possibly my head.

I was so scared that my memories are indistinct and covered by amnesia. Even as I write this, I am aware that a great deal more happened. I just can't get to it. This might be terror amnesia, or drugs, or hypnosis, or even doses of all three. There is one drug, tetradotoxin, which could approximate such a state. In small doses it causes external anesthesia. Larger doses bring about the "out of the body" sensation occasionally reported by victims of visitor abduction. Greater quantities can cause the appearance of death—even the brain ceases detectable function.

This rare drug is the core of the zombie poison of Haiti, and little is known about why it works as it does. It is also the notorious "fugu" poison of Japan, found in the tissues of a blowfish, which is an esteemed if deadly aphrodisiac.

My surroundings were so unfamiliar in every detail and my surprise was so great that I simply faded away, in the sense that my ability to direct myself was lost, mentally as well as physically. Not only was I physically anesthesized (although no longer so much paralyzed as totally limp), I was in a mental state that separated me from myself so completely that I had no way to filter my emotions or most immediate reactions, nor could my personality initiate anything. I was reduced to raw biological response. It was as if my forebrain had been separated from the rest of my system, and all that remained was a primitive creature, in effect the ape out of which we evolved long ago.

I was not, however, in the ape. I was in my forebrain, locked away from the rest of myself. My mind had become a prison.

One being was on my right, another on my left. Within my field of vision a great deal of rushing about commenced again. The next thing I knew, I was being shown a tiny gray box with a sliding lid. There was a curved lip at one end of this box, to make it easy to push it open. It was being held by a thin, graceful person whose appearance was not distinct. Was this the female again? I'm not sure. It almost seems, as I remember, that something had been done to my eyes to affect my ability to concentrate my vision. Glances around the room were quite detailed in recollection, but any attempts to steady my vision and view a particular being resulted in blurring. It would be interesting to know if this was an induced effect or something caused by my own fear of what I was seeing.

My memory of the one that came before me next is of a tiny, squat person, crouching as if huddled over something. He had been given the box and now slid it open, revealing an extremely shiny, hair-thin needle mounted on a black surface. This needle glittered when I saw it out of the corner of my eye, but was practically invisible straight on.

I became aware—I think I was told—that they proposed to insert this into my brain.

If I had been afraid before, I now became quite simply crazed with terror. I argued with them. "This place is filthy," I remember saying. Then, "You'll ruin a beautiful mind." I could imagine my family awakening in the morning and finding me a vegetable. A great sadness overtook me. I do not recall screaming, but evidently I was doing so, because I remember the next exchange quite clearly.

One of them, I think it was the one I had identified earlier as the woman, said, "What can we do to help you stop screaming?" This voice was remarkable. It was definitely aural, that is to say, I heard it rather than sensed it. It had a subtly electronic tone to it, the accents flat and startlingly Midwestern.

My reply was unexpected. I heard myself say, "You could let me smell you." I was embarrassed; that is not a normal request, and it bothered me. But it made a great deal of sense, as I have afterward realized.

The one to my right replied, "Oh, OK, I can do that," in a similar voice, speaking very rapidly, and held his hand against my face, cradling my head with his other hand. The odor was distinct, and gave me exactly what I needed, an anchor in reality. It remained the most convincing aspect of the whole memory, because that odor was completely indistinguishable from a real one. It did not seem in any way a dream experience or a hallucination. I remembered it as an actual smell.

There was a slight scent of cardboard to it, as if the sleeve of the coverall that was partly pressed against my face were made of some substance like paper. The hand itself had a faint but distinctly organic sourness in its odor. It was not a human smell, but it was unmistakably the smell of something alive. There was a subtle overtone that seemed a little like cinnamon.

The next thing I knew, there was a bang and a flash, and I realized that they had performed the proposed operation on my head. I felt like weeping and I recall sinking down into a cradle of tiny arms.

At this point, I had some feeling, and enough muscle tone had returned to enable me to slide my feet along the floor in an effort to avoid falling all the way. Then I was lifted up and seemed suddenly to be in another room, or perhaps I simply saw my present surroundings differently. It appeared to be a small operating theater. I was in the center of it on a table, and three tiers of benches were populated with a few huddled figures, some with round, as opposed to slanted, eyes.

I was aware that I had seen four different types of figures. The first was the small robotlike being that had led the way into my bedroom. He was followed by a large group of short, stocky ones in the dark-blue coveralls. These had wide faces, appearing either dark gray or dark blue in that light, with glittering deep-set eyes, pug noses, and broad, somewhat human mouths. Inside the room, I encountered two types of creature that did not look at all human. The most provocative of these was about five feet tall, very slender and delicate, with extremely prominent and mesmerizing black slanted eyes. This being had an almost vestigial mouth and nose. The huddled figures in the theater were somewhat smaller, with similarly shaped heads but round, black eyes like large buttons.

Throughout the whole experience, the stocky ones were always present. They were apparently responsible for moving and controlling me, and I had the distinct impression that they were a sort of "good army." Why good I do not know.

I do not remember what, if anything, happened in the operating theater. My memories of movement from place to place are the hardest to recall because it was then that I felt the most helpless. My fear would rise when they touched me. Their hands were soft, even soothing, but there were so many of them that it felt a little as if I were being passed along by rows of insects. It was very distressing.

Soon I was in more intimate surroundings once again. There were clothes strewn about, and two of the stocky ones drew my legs apart. The next thing I knew I was being shown an enormous and extremely ugly object, gray and scaly, with a sort of network of wires on the end. It was at least a foot long, narrow, and triangular in structure. They inserted this thing into my rectum. It seemed to swarm into me as if it had a life of its own. Apparently its purpose was to take samples, possibly of fecal matter, but at the time I had the impression that I was being raped, and for the first time I felt anger.

Only when the thing was withdrawn did I see that it was a mechanical device. The individual holding it pointed to the wire cage on the tip and seemed to warn me about something. But what? I never found out.

Events once again started moving very quickly.

One of them took my right hand and made an incision on my forefinger. There was no pain at all. Abruptly, my memories end. There isn't even blackness, just morning.

I had no further recollection of the incident.

I awoke the morning of the twenty-seventh very much as usual, but grappling with a distinct sense of unease and a very improbable but intense memory of seeing a barn owl staring at me through the window sometime during the night.

I remember how I felt in the gathering evening of the twenty-seventh, when I looked out onto the roof and saw that there were no owl tracks in the snow. I knew I had not seen an owl. I shuddered, suddenly cold, and drew back from the window, withdrawing from the night that was falling so swiftly in the woods beyond.

But I wanted desperately to believe in that owl. I told my wife about it. She was polite, but commented about the absence of tracks. I really very much wanted to convince her of it, though. Even more, I wanted to convince myself. So intent was I on this that I telephoned a friend in California for the specific, yet unlikely, purpose of telling her about the barn owl at the window.

Later I discovered that memories of animals in strange places are a common block to this experience. One young woman arrived back at a picnic in the woods in France with a story of seeing a beautiful deer. But she had blood on her blouse, and a strange straight scar that could not be explained. Ten years passed before she remembered anything of the truth of her experience in those woods, and she would have died with that memory had not her memory of another encounter with the visitors caused her to question its real significance. Another man came away from his experience thinking only that he had seen a bunch of rabbits hopping around outside his car.

Like my barn owl, these stories must have seemed no more than whimsies, but they hid real experiences that were so impossible to accept, just keeping them hidden took a large toll—as it has with others, as it might be doing with anybody.

From that first day my wife noticed a dramatic personality change in me, which she thought was similar to a change that had taken place the previous October. We had gone through personal hell then because of my demands and accusatory behavior, and she did not want that pattern to repeat itself.

But I was in decline again, and this time the symptoms were not all mental. That first evening I underwent the initial physical symptom of my ordeal. We had come in from an afternoon of light cross-country skiing, not at all strenuous. I was dead tired. Normally I am full of energy. Even a hard afternoon on the ski trails leaves me feeling pleasantly relaxed.

I got chills and went to bed. I lay huddled between the sheets and the quilt, with evening coming down, feeling just awful. I thought that I must have had a high fever. I was exhausted. The sounds of my wife and son downstairs filled me with a sense of foreboding. Strange recollections of people running, of being pulled and shoved, swirled through my head.

Then our nearest neighbors suddenly arrived. They appeared without warning. We tend to be very private in our sparse community, and this was only their second spontaneous visit in the two years we have been neighbors.

Feeling somewhat better, I went downstairs to see them. No sooner had we started talking than I found myself complaining that I thought I had seen the light of a snowmobile in the woods between our houses at about three in the morning. I was horrified at myself. What was I saying? I couldn't remember any such thing, and I knew it even as I spoke. Our neighbors offered the thought that the woods were too thick for a snowmobile to maneuver, which is true. Then I said that it must have been the lights on his house. He has two floodlights that shine out over his backyard. He explained that these lights had been off, but promised to redirect them so they couldn't be seen from our house. I knew even then that his lights hadn't been bothering me so late at night (although they were sometimes bothersome early in the evening, now that winter had stripped the woods of their concealing leaves). My memory of the snowmobile was as hollow as my memory of the owl.

After some small talk, our neighbors went home. I was not pleased with my own behavior, and found it hard to understand because it seemed so nonvolitional, almost as if I had been talking against my will.

My wife reports that my personality deteriorated dramatically over the following weeks. I became hypersensitive, easily confused, and, worst of all, short with my son. We have always been a happy family, and there was no change in our life condition or relationship to account for this personality shift.

The realization that the owl memory was not true created troubling problems for me. I was aware that something had somehow gone wrong with me. The trouble was I could not understand what it was. There simply wasn't anything in my life to explain it. I started to worry about toxins in our food or water, but as nobody else in the family was affected, and we hadn't tried any food that might have caused some bizarre allergy, that seemed unlikely.

I did not know that the owl and the light were screen memories that concealed a traumatic experience. As described by Freud, the screen memory is a method that the mind uses to shield itself from things too upsetting to recall.

I had a feeling of being separated from myself, as if either I was unreal or the world around me was unreal. By December 28 I was so depressed and in such a state of inner conflict that I sat down and wrote a short story in an effort to explore my emotions. It reflected not only my emotional state but probably also some of the realities hidden behind it. I called it "Pain." It was to be the last sustained writing I would do for seven weeks, and is the last thing I wrote before these enormous hidden truths began emerging from my unconscious.

At the time I had no idea that I was suffering from emotional trauma, or that dozens of other people had been through very similar ordeals after being taken by the visitors.

Previous to the twenty-sixth I had made good progress on a huge two-volume novel based on the relationship between Russia and America at the outset of the Russian Revolution. Now I could hardly read my own work, let alone continue working on this complicated project.

Since I wrote "Pain" as an expression of the emotional state that overlay the memories I was then suppressing, I will recount it briefly. It is about a man who encounters an enigmatic woman named Janet, who proves to be some sort of superhuman being, perhaps an angel or a demon. She draws this man into a strange experience of capture and incarceration in a tiny, magical cabinet. From the agony that ensues, he gains immense insight and new spiritual strength.

What is most interesting to me about this story is that it continues imagery that is present in my early horror novels. The visitors could be seen as the Wolfen, as Miriam Blaylock in *The Hunger,* and as the fairy queen Leannan and her soldiers in *Catmagic.* The theme is always the same: Mankind must face a harsh but enigmatically beautiful force that, as Miriam Blaylock describes herself, is "part of the justice of the world." This force is always hidden between the folds of experience.

As I worked on "Pain" my mental and physical states continued to get worse. An infection appeared on my right forefinger. It looked like a splinter injury, but I could not remember getting a splinter, unless it was from some log I carried in for the stove. The injury festered. Neither iodine nor antibiotic ointment seemed to help. I looked for a bit of splinter but could find nothing.

I noticed that I was uncomfortable sitting because of rectal pain, a weird and disturbing symptom. I had a vague feeling that something distressful had happened to me, but no clear memory.

In the ensuing days, I experienced more bouts of fatigue. I would be working and suddenly I would get cold and start to shake. Then I would feel so exhausted that I could not go on, and crawl into bed quivering and miserable, sure that I was coming down with the flu. I took my temperature during one of these experiences and found that it was 96.6 at the outset and 98.8 at the height of the "fever." Afterward it dropped to 97.0.

Nights I would sleep, but wake up in the morning feeling as if I had been tossing and turning the whole time. I ceased to dream, and sometimes had difficulty closing my eyes. I felt watched, and kept hearing noises in the night.

Mornings I would wake up with the feeling that I had been somehow on guard.

My disposition got worse. I became mercurial, frantic with excitement about some idea one moment, in despair the next. I was suspicious of friends and family, often openly hostile. I came to hate telephone calls. I could not concentrate even on light television programs. After writing "Pain" I found that I could not sustain enough attention to work for more than five or ten minutes at a time. An attempt to read *Gerald's Party* by Robert Coover left me profoundly confused. I kept reading and rereading the same few pages. I switched to a less challenging novel, but it was also totally incomprehensible. I had been reading some sermons of the thirteenth-century mystical philosopher Meister Eckhart, but this study had to be abandoned. I could no longer follow my own thinking, let alone that of the authors who interested me. It was a fearful, haunting discovery.

On the afternoon of January 3 we were skiing when I got a pain behind my right ear. It was a sensation similar to what happens to one's jaw when Novocain wears off after a session in the dentist's chair. My skull ached and the skin was sensitive. In the middle of this sensitive area my wife could see a tiny pinpoint of a scab.

I believe that the combination of the infected finger, the rectal pain, and the aching head were what finally brought my memories into focus. The confused swirl resolved into a specific series of recollections, and when I saw what they were, I just about exploded with terror and utter disbelief.

One of the memories would come into my head, linger there a moment, then leave me with my heart pounding, gasping, sweat pouring down my face.

I thought that I had lost my mind.

For half of my life I have been engaged in a rigorous and detailed search for a finer state of consciousness. Now I thought my mind was turning against me, that my years of eager study of everything from Zen to quantum physics had led me into some strange and tragic byway of the soul.

As soon as I had them in focus, my memories became perfectly vivid—as vivid, say, as childhood memories become when one chooses to draw them out of the mental file where they are hidden. I sat at my desk, trying to make sense of what could not make sense.

I thought, quietly and calmly, *You may be going mad, or you may have a brain tumor. You've got to find out which it is and take whatever steps are necessary.* And then I rested my head on the desk, and, quite frankly, cried.

For a couple of days I lived with it. Maybe the "symptoms" would subside.

Then, quite suddenly one afternoon, I recalled the smell. Their smell. It came back to me as clearly as if I had inhaled it not a moment before. More than anything except discovering that I was not alone with my experience, that totally real memory saved me from going stark raving mad.

In the first week of January, a local newspaper published accounts of an object or objects being sighted in our area. This story appeared in the January 3, 1986, issue of the Middletown, New York, *Record*. The headline called the appearance a hoax, but according to the story local people who had witnessed the event doubted that. One man, however, claimed that he had seen the things fly over a brightly lit local prison, and in the light he saw planes. A follow-up story on January 12 expanded on the prank hypothesis.

My wife showed me the article and told me, "You said this would happen. You were talking about this last week." I did not remember the conversation, but the article caused me to glance over a book my brother had sent me for Christmas called *Science and the UFOs* by Jenny Randles and the astronomer Peter Warrington.

Warrington is a respected scientist, and the book seemed well written. As a matter of fact, it does not make any claims about the reality of the phenomenon, but simply calls for more study and appeals to the scientific community to begin to accept it as a legitimate area of inquiry.

I was surprised to find that *Science and the UFOs* frightened me. I put it aside with no more than the first five or six pages read.

Much later, after we had really begun to take this whole matter seriously, Anne and I did more research into UFO sightings in our area. We discovered that it is a hotbed of sightings, and has been for nearly half a century.

As it happens, the eighteen-year-old son of one of our neighbors saw something hovering near a road not five miles from our cabin at approximately nine-thirty on a night in late December. He described it to me as "huge and covered with lights," a typical description. He watched it for some time. Being the son of a former state trooper and pilot, he did not claim that it was a "UFO," but simply told the truth: He did not know what it was, but it appeared to be a solid structure, and as it hovered for a substantial period, more than fifteen minutes, it could not have been a flight of planes. I telephoned the Goodyear Corporation and found that their blimp was not in the area at the time.

The only thing I thought it could have been was some unknown blimp, but even that appeared hard to believe in view of what more I discovered about it.

Just by talking with friends in April my wife uncovered a personal experience of an early area sighting, one that took place in the late fifties. One of her best friends is an artist and the wife of a well-known composer. In her childhood this woman used to attend summer camp at a location not ten miles from where our log cabin now stands, a fact that we did not know when Anne asked her the question we had determined to put to as many people as we could, as part of our research effort.

To at once gain valid information and prevent bias, we had simply been asking people, "What was your strangest experience?" None of the people we asked had any idea of what was happening to us.

The woman's answer turned out to be highly revealing. She reported that she had seen a flying saucer in 1953, when she was nine. She proceeded to describe an object similar to the one that had appeared in the same immediate area in December 1985. Like so many reports through the years, she described it as huge, full of lights, and hovering. It moved off slowly.

If all these objects were the results of pranks, then the pranksters would have been operating for more than thirty years—and even in the early fifties they would have had superb mufflers, considering that the object seen then made no more sound than the ones seen today.

Further research revealed to us that our area of upstate New York, comprising roughly Westchester, Orange, Putnam, Rockland, and Ulster counties, had an absolutely extraordinary series of sightings of boomerang- or triangular-shaped objects of enormous size, starting in 1983. Thousands of people saw these objects, ranging from meteorologists and Federal Aviation Administration employees to a whole cross section of local people. Town police officers, sheriffs, state troopers, even in one case an entire town government en masse viewed the things, which have been described as being "the size of an aircraft carrier."

The official explanation, detailed in *Discover* magazine in November 1984, was that the sightings were created by a group of pilots flying light aircraft. According to *Discover*, the light aircraft sometimes flew in formation with their engines off and their wing tips six inches apart at night. Since these planes never seemed to use their radios, it was subsequently added in local newspapers that radio silence was maintained during these tight nighttime maneuvers.

A pilot told me that this was all highly unlikely, that such formation flying would not be possible even with much heavier aircraft. Pranksters and even secret aircraft may be part of the answer to the enigma, but they are not the whole answer.

An article appeared in the April 17, 1983, issue of *The New York Times* quoting a professional meteorologist who observed a silent object a thousand yards in diameter hovering a hundred feet above him. He is quoted by the *Times* as saying that he had the sensation of "being scanned and rejected."

I do not think that a professional meteorologist would mistake an object nearly a mile wide for a flight of airplanes, not at an altitude of a hundred feet.

Mr. Philip J. Klass, a noted debunker of unexplained UFO sightings, claimed that people were probably seeing "an advertising airplane." Mr. Klass was at that time an editor of *Aviation Week and Space Technology*, a publication noted for its uncanny ability to obtain scoops from the Department of Defense about secret aerospace projects. Mr. Klass also writes for a publication I have admired, *The Skeptical Inquirer*. In view of my own experiences, however, I am beginning to suspect that, in the case of this particular chimera, skepticism has been taken farther than is reasonable or wise.

Neither the official story nor Mr. Klass's offering explains the hundreds of closeup sighting reports collected by local science teacher Phillip J. Imbrogno, whom the *Times* described as "one person working hard to provide a rational explanation." I spoke to Mr. Imbrogno, who said that he had collected since 1983 more than two hundred reports from people trained in some way as observers, and that they had seen huge devices that had clear structure to them. He added that on one night when there were extensive and clear sightings of a device hovering above a local parkway, the winds were averaging 23 knots! What people saw on that night was not aircraft, heavy or light, flying in close formation.

And nobody has explained who came and took me in the night and injected something into my brain.

When we went to New York City for a stay in January we still knew almost nothing about UFOs, and nothing at all about the sightings discussed above.

Life did not return to normal. Even though there was no further reason for me to delay writing, I couldn't seem to get down to work. I felt a little better, but I was so terribly uneasy. My difficulty relating to my wife and son continued.

I finally finished *Science and the UFOs*. Toward the end of the book I was astonished to read a description of an experience similar to my own. When I read the author's version of the "archetypal abduction experience," I was shocked. I was lying in bed at the time, and I just stared and stared at the words. I, also, had been seated in a little depression in the woods. And I had later remembered an animal.

My first reaction was to slam the book closed as if it contained a coiled snake.

They were talking about people who think they're taken aboard spaceships by aliens. And I seemed to be such a person. My blood went cold: Nobody must ever, ever know about this, not even Anne. I decided just to lock the business away in my mind.

A few mornings later at about ten, I was sitting at my desk when things just seemed to cave in on me. Wave after wave of sorrow passed over me. I looked at the window with hunger. I wanted to jump. I wanted to die. I just could not bear this memory, and I could not get rid of it. What on earth *were* those things? What had they done to me? Were they real, or was I the victim of some unknown mental state?

I remembered that a man named Budd Hopkins had been mentioned in the book as a prominent researcher in the field. The name had been familiar to me: Anne and I are interested in art and Hopkins is a well-known abstract artist, collected by the Guggenheim and the Whitney.

I found his name in the phone book. But how could I call him? What a stupid thing to have to admit. Little men. Flying saucers. How idiotic.

I recognized clearly, though, that if I had another moment of despair that intense, I was going to go out the window. No question. I owed it to the family who loved and depended on me to try to help myself.

I called Budd Hopkins. He answered the phone and listened to my story for a few minutes. I thought I would wither away with embarrassment telling it, but he soon interrupted me. Could I come over—like right now?

It turned out that his place in the city was a ten-minute walk from mine.

Hopkins was a large, intense man with one of the kindest faces I have ever seen. I later discovered that he was bright and canny, but at the time he assumed a guileless appearance.

The moment our interview began, Hopkins explained that he was not a therapist but he could put me in touch with one if I wanted that. He then got the facts from me that I have recorded here.

As I sat there in that man's living room, listening to him tell me that I wasn't alone, that others had gone through very much the same thing, the tears rolled down my cheeks, and I went from wanting to hide it all to wanting to understand it.

It was during that first meeting that he asked me if anything else had happened in the past, anything unusual. My initial reaction was to say no. One of these ludicrous and horrible experiences was quite enough. But the question seemed to trigger something in me. After a moment's reflection, I blurted out, "I seem to remember a night the house burned down. But it didn't burn down."

All hell had broken loose on a night in early October. There had even been an explosion that woke up the whole household. Strange things had happened, but for some unknown reason we had simply put them out of mind. We'd hardly even discussed them. But that time they didn't happen only to me. We'd had houseguests. If anything had really happened, they would certainly remember. Here was a chance to put this to the test. If nobody remembered anything, I would be able to dismiss this embarrassing business of aliens.

I left Hopkins's house a happy man. He'd said that judging from his experience, the October events may have been caused by the same agency that was responsible for the disturbance of December 26.

Wonderful. I'd contact the witnesses. They would of course report that they remembered nothing. Then I would begin the painful but thankfully well-understood process of accepting that I had been the victim of some unusual mental phenomenon. I would enter therapy and learn to forget the mysterious visitors.

No other outcome seemed possible. Of course not. I had solved my problem.

October 4, 1985

As I walked down Sixth Avenue toward my end of Green-wich Village, what had happened on October 4 became clearer in my mind.

Why had I not thought of or discussed these events before? The answer is straightforward: I had tucked the whole episode into the catalog of open questions and forgotten it. In retrospect the only reason I can advance for having done this is that I did not want to face just how strange the events of that night had really been. But when I thought them over, they began to seem distinctly eerie, even frightening.

We often deal with fear by rejection—and in this case, as will soon be evident, there was more than enough reason to be terrified.

When I wrote the narrative that follows I had not yet been hypnotized and did not know what, if anything, lay unseen in my mind. I wrote it over a two-day period after first seeing Hopkins and sometime before I met Dr. Donald Klein, who would become my hypnotist when we began to discover empty places in my memory.

On October 4, 1985, my wife, son, and I drove up to our cabin in the company of two close friends, Jacques Sandulescu and Annie Gottlieb.

We have known Jacques and Annie for about five years. The thing about them most immediately apparent is that he is as enormous as she is tiny. He weighs nearly 300 pounds, is a black belt in karate, and does a hundred push-ups at a session. She is also a black belt, but weighs perhaps 120 pounds. She is intellectual, he is physical. Both are writers. He came to the United States as a refugee from slave labor in the Soviet Union in the late forties. A Rumanian na-tional, he had been forcibly transferred to the Donbas region to be worked to death in the mines there. His book, *Donbas,* tells of his long journey of escape, and paints an accurate picture of him as a profoundly physical man. He would make a good witness, I thought, because of his steadfast sense of reality.

Annie Gottlieb is more an intellectual, the author of the recent *Do You Believe in Magic: The Second Coming of the Sixties Generation* (Times Books, 1987).

The night of October fourth was foggy in Ulster County. We had dinner at a local restaurant and arrived at the cabin at about nine in the evening. I turned on the pool heater so that the pool would be comfortable for use the next day (Saturday). Then I lit a fire in the wood stove. We were all sleepy, so sleepy that we went off to bed almost immediately.

Anne and I retired to our upstairs bedroom, Jacques and Annie went to the guest room and closed the door, and our son went to his corner bedroom beside theirs. He left his door open. From my bed, with the bedroom doors open, I could see out across the cathedral ceiling of the living room to a hexagonal window set in the peak of the roof.

Over the next hour, the fog grew thicker and thicker. When I turned out my reading light I was enveloped in absolute blackness and total silence. The harvest moon had been full on September 29, and was now at about half. It rose at approximately ten-thirty, but was entirely invisible because of the cloud cover.

I do not remember what I had been reading that night, but it wasn't frightening, nor was dinner the sort of meal that would give rise to later unrest. We had not drunk more than one glass of wine and a drink each at the restaurant.

I slept dreamlessly for some period of time, perhaps as much as two or three hours. Then I was startled awake and saw, to my horror, that there was a distinct blue light being cast on the living-room ceiling.

I was frightened, because it wasn't possible for there to be any light there. Car lights from the road could not be cast on that ceiling. In early October our neighbor was away in Japan, and his house was not only dark but invisible through the forest between our places, which was still thickly leafed. The automatic porch light that had been persistently troublesome was now without bulbs. It could not have been a flashlight, because it was so uniform and so broad, and so distinctly blue. We have tried duplicating the light with a fluorescent camp lantern both on a clear night and on a similarly foggy one, but even an extremely powerful fluorescent light could not achieve the effect, let alone a small portable unit.

My mind inventoried the possibilities as I watched this blue light slowly creep up the ceiling, as if whatever was causing it were slowly moving down into the front yard from above. Finally I hit on what seemed to me a sensible solution: The chimney must be on fire and dropping sparks into the front yard. I had to do something about it at once.

Then I fell into a deep sleep! The last thought I remember before dropping off with my heart still hammering was that the roof was on fire. This was the first such wildly inappropriate reaction on that night, but it was not to be the last.

I do not know what time this all took place, but it was well after midnight.

Sometime after I fell back to sleep I was again awakened, this time by a loud report, as if a firecracker had popped in my face. My wife cried out and downstairs my son began shouting.

When I opened my eyes I was stunned to see that the entire house was surrounded by a glow that extended into the fog.

I thought to myself: *You damned fool, you fell asleep and now the fire's gotten worse*. I finally managed to get up. As I did so I said to Anne: "The roof's on fire. I'll get our son and wake up the others." I started downstairs.

I hadn't gotten halfway across the room before the glow suddenly disappeared. I was very confused. There was nothing to do but tell Anne that I had made a mistake, then go downstairs to comfort my son. On the way I met

Jacques in the hall. His presence terrified me and I jumped back away from him. Then I apologized for being so startled by a friend, told him to calm down and go to bed, and added that nothing was wrong.

I continued into my son's room and embraced him. In a few minutes I was back in bed and the household was again asleep.

The next morning little was said about the incident. I do remember Annie mentioning that Jacques had been bothered by the light the night before. I didn't understand that because their bedroom door had been closed, so they couldn't have seen the bathroom light, which is left on for our son. I didn't remember my confusion about fire. As far as I was concerned, Annie and Jacques had been disturbed by the light but I hadn't been.

Later that week I found myself a little agitated, without knowing quite why. I had a persistent memory of light flashing in my eyes that night. And I vaguely recalled some sort of an explosion.

The next weekend I had a very clear and dramatic memory of a huge crystal standing on end above the house, a glorious thing hundreds of feet tall, glowing with unearthly blue light.

I told Anne about it, and as I was talking I experienced a hollow sort of a feeling. I knew that she didn't believe me— of course she didn't! And I didn't believe myself. "Wasn't there some problem with the stove?" she asked. I was embarrassed and never mentioned the crystal again. I put it out of my mind permanently.

On February 6, 1986, I came home from Hopkins's house brimming with eagerness. I was sure I would put an end to this by asking careful questions. Jacques and Annie *had* been disturbed by the bathroom light. Of course. Their door must have been opened, as I had seen Jacques in the hall. And my Anne had cried out not because of the explosion but because I had told her the house was on fire. There had been no explosion. And as for the blue light on the living-room ceiling, put some unanticipated light source together with thick fog and anything can happen.

I first asked my wife to think back to October 4. It wasn't hard to identify the specific night, because it was the last time Jacques and Annie had come to the country and the thickness of the fog was unusual.

I was disturbed that Anne at once remembered being awakened by the bang. She did not see the glow, but my initial warning about the fire apparently didn't penetrate her sleep, because all she did recall was my saying that there was no fire.

I asked my son, "Do you remember the last time Jacques and Annie went to the country with us?"

"Yeah. The night of the bang." So he had also heard it. "A bunch of people told me it was OK; you just threw your shoe at a fly."

"What people?"

"Just a bunch of people. People who were around."

This answer, I must admit, shocked me badly. I left off questioning him and called Budd Hopkins, who suggested that I ask my son not about memories but about dreams.

Taking this advice, I next asked my son if he remembered any unusual dreams. This is his reply, spontaneous and immediate:

"I dreamed that a bunch of little doctors took me out on the porch and put me on a cot. I got scared and they started saying 'We won't hurt you' over and over in my head. That is my strangest dream, because it was just like it was real. It happened in the middle of another dream, when I was dreaming that me and Ezra [a friend of his] were in a boat." He could not say if he had had the dream on the night of October 4. He knew only that it had happened at the cabin.

His words swept away all my hopes of solving this problem in anything remotely resembling a conventional manner. What had happened to my little boy? His innocent report was very upsetting. In the context of my own experiences, his dream suggested either that the two of us have some sort of weird psychological link, or that at some point he has had an experience similar to my own.

Next I spoke to Jacques Sandulescu on the phone. This is a transcript of that conversation.

Me: "Do you remember anything about the last time you and Annie came to the country?"

Jacques: "The light! I was sleeping, all of a sudden something woke me up. I saw the room was full of light. Bright, like daylight. Not like the moon. I thought we overslept. I look at my watch, it says four-thirty. Then I hear you through the door, saying it's OK. The light is gone, so I go back to sleep."

Me: "What kind of light was it?"

Jacques: "Light, it was light. I could see the bushes outside. I could see the tree trunks. I thought it was about ten in the morning."

I have done every conceivable thing to try to duplicate light like that. Our guest room has one small window overlooking a seven-feet-deep covered porch. Beyond that the land slopes up gradually, so that not even car lights from the road can enter that room, much less moonlight or sunlight. With the leaves gone during the winter, we determined that the lights from the neighbor's house are also invisible from that window. The movement-sensitive light doesn't shine directly in, but down the porch. Had it somehow turned on—even absent bulbs—Jacques would have seen not the trees and shrubs but the outline of the porch interior with the yard beyond in darkness. The reason for this is that the light shines past the window and down the length of the porch. Had the regular porch light been switched on, the same effect would have resulted.

Even with the neighbor's lights on, the porch light on, and a car in the front yard, we could not duplicate the effect. Nothing I can conceive of can account for the major light phenomena on that night. It may be possible to explain the blue glow I originally saw on the ceiling, but not that massive burst of light from above. I visualized the whole roof being ablaze. Jacques thought it was midmorning. Because of the fog, a helicopter, or indeed any sort of airplane, was out of the question. A pilot told me simply, "Forget aircraft."

At four-thirty the moon was still in the sky, but well below the line of the forest. Could the fog have somehow magnified the moonlight, causing dark-sensitized eyes to mistake its mild glow for bright daylight? Such a thing may have been possible, but the moon was low in the west and the source of the light was clearly directly overhead. And what about the explosion? Maybe it was thunder. But there were no thunderstorms in the area. Perhaps a freak bolt in some sort of unusual ministorm caused it, then. But the period of seeming daylight lasted many seconds, and was not apparent to anybody until *after* the explosion. Thunder follows lightning, not the other way around.

Whatever caused the effect, it was a highly unusual phenomenon and it is unlikely that it can be identified.

And so far there is no way at all to account for Annie Gottlieb's testimony. I spoke to her immediately after talking to Jacques. While she must have overheard him on the phone, the two of them had no time between statements to discuss the matter. Also, they are normal, coherent, and reliable people. They had, and have, no reason whatsoever to lie and they are most unlikely to be so radically confused by normal realities that they would derive from them memories such as they report. One only has to look into Annie Gottlieb's writings to see the clarity of her mind.

Like the rest of us, Annie was awakened by a loud explosion. She reports: "It was a bang. Then I heard the scurry of little feet running across your bedroom upstairs. It must have been the cats."

"Annie, the cats were in the city. We don't take them weekends because they don't like the carrier."

"You're kidding! I always just assumed it was the cats. Anyway, I vaguely remember the light. Mostly I remember the noises. A few minutes after the scurrying, I heard you come downstairs. You said through the door not to worry. The next morning you told me that some people had come down from a spaceship to visit."

"*What?* Annie, I never said any such thing. I would *never* say anything like that."

"At the time I thought it must have been some kind of dream."

"You remember me saying it?"

"Well, now that I think of it, I don't know where I got the idea that anybody said that." (Months later she recalled that I had not spoken about a visit, but had described the crystal. In any case, I certainly had a very strange explanation for the night's disturbances.)

At that point I almost wished that I had never asked my witnesses anything. I said good-bye and put the phone down. I realized, finally and inescapably, that something very peculiar was going on. I could not deny it. I would have been a fool to deny it.

I went into my office and closed my door. It was evening, and Manhattan's few blind stars were shining in the sky. The world outside looked so normal, and that moment its very normality seemed to me to be the most beautiful thing I had ever seen.

I thought back across the months to October. The fall had been an awful time for me and my wife. Around the second week of October I had become extremely fearful about living in the New York area and decided to move.

Had my terror stemmed from that night? And what about all my nervousness, my secret searching under beds and in closets, my unreasonable fear of prowlers? It seemed to me that I had been growing increasingly uneasy with the passing months. I had awful dreams that I cannot remember. Again and again I woke up in the small hours of the morning feeling as if something dreadful had just happened.

The last week of October, still with no conscious memory of that night being in any way unusual, I decided that I couldn't live a moment longer in New York. The city streets seemed hideously dangerous. Our cabin was a dark, terrible place, one that I could not bear ever to enter again. I felt out of control, as if anything could happen, and might.

I decided that I wanted to move to Austin. I went to the University of Texas there, and it is a city that both Anne and I love. Some of our best friends, including my collaborator James Kunetka, live there.

I insisted on putting both the cabin and the apartment on the market.

After Halloween we went down to Texas and arranged for our son to attend a local private school and began the process of buying a house.

We got an offer on our cabin, but no interest in the apartment.

One evening in Austin we were looking at the house we had chosen to buy. My wife was inside talking to the realtor and the owners. I walked out onto the deck.

When I looked at the dark canyon that stretched out into the shadows, and the stars in the evening sky, I felt suddenly and absolutely afraid. It was exactly as if the sky were a living thing, and it was watching me.

What was even more frightening was my clear awareness that this was a paranoid fantasy. I thought then that my mental health was not good, and soon I would either have to calm down or take steps to improve it.

But I could not live in that house. In fact, I could never enter it again.

When I changed my mind and decided to stay in New York, my wife was understandably furious. Then I accused her of being the one who had wanted to move us to Austin.

There followed a crisis. She really thought that she might have to leave me, because life together was just getting intolerable. But we are a deep marriage, and her despairing threat to separate made me quell my extreme behavior. It was not until Christmas that I really began to feel better.

Sitting in my office that afternoon in February, I took stock of all I had found out. I had promised Hopkins that I wouldn't read anything about unidentified flying objects. In the past, as I have said, my interest in the subject was minimal. I have certainly read a book or two about them. Pressing myself I thought maybe I could remember seeing something years ago in *Look* magazine about somebody named Hill being taken aboard a flying disk. (In July 1986 I got copies of the issues involved—October 4 and 18, 1966—and I do not think that I actually read them at the time. I must have seen something about the story, though, because I remember it. Maybe there was a report in the newspaper.)

Judging from what the other witnesses reported, something had happened. But what? Even after talking to Hopkins, I was by no means willing to ascribe my experiences to the UFO phenomenon. I wanted to be quite clear: I had no idea what had gone on that night. There did seem to be a lot of confusion, though, and perhaps even an emotional response on my part greatly out of proportion to what seemed a minor disturbance.

TWO

O plunge your hands in water,
Plunge them in up to the wrist;
Stare, stare in the basin
And wonder what you've missed.

The glacier knocks in the cupboard,
The desert sighs in the bed,
And a crack in the tea cup opens
A lane to the land of the dead.
—W. H. AUDEN, "As I Walked
Out One Evening"

DOWN THE CAVE OF MIND

Hypnosis

The Uncertain Mirror

My next step was clear. I was going to become involved with a therapist. But I had certain criteria. It could not be somebody who believed anything in particular about visitors or the disk phenomenon. The ideal therapist would have an open mind: I could have a mental problem. It might or might not have components unknown to science. Or it could be just what it seemed.

Because of the evident presence of fear-induced memory lapses and even possible amnesia, this therapist would have to be a skilled hypnotist as well. And again, not just any psychiatrist using hypnosis in his practice would do. I wanted somebody with a reputation in the scientific community as a real expert. I wanted both scientific rigor and therapeutic skill—and the two are not always present in the same person.

I chose not to approach any hypnotist to whom Hopkins had made previous referrals, despite the excellence of their credentials. One of these, Dr. Aphrodite Clamar, had worked extensively with Hopkins and was a very fine and highly professional psychologist, but I was firm in my desire to do this with somebody who had had no previous involvement.

Hopkins remembered that Dr. Donald Klein of the New York State Psychiatric Institute had expressed interest in the phenomenon and appeared to be open-minded about it. I looked up Dr. Klein's credentials and found them to be superb. If he would take me on, he was the ideal man.

A few weeks later I was in his office undergoing a searching three-hour preinterview. I had provided him with a document outlining all my memories. We worked for some time trying to find ways into my mind, but I could recall little more than I already had. At his suggestion, I spent a week trying to do so. When I was not successful—in fact, all I got out of it was dizziness and strange nightmares—we decided on a trial hypnosis session.

I was dubious about hypnosis. I'd read in *Science News* of a study that suggested that anybody under hypnosis can be induced to remember a "UFO abduction and experience," complete with little men and all the trimmings. The hypnotist has only to ask the right questions, and the stories apparently just come pouring out.

It is a flat-out myth that people can't lie under hypnosis. They can and they will—if they think that's what the hypnotist wants them to do, or if they themselves want to do it.

When I got a more careful look at the study I had read about in *Science News,* I found that the questions asked were intentionally leading ones, specifically designed to evoke abduction memories. This study had as its purpose to prove that anybody can be induced to relate an abduction experience under hypnosis if he or she is asked questions designed to suggest that the hypnotist wanted him or her to relate such an experience.

Well, there was no chance at all that Dr. Klein was going to do that. This can easily be confirmed by the reader, as all the transcripts of my hypnosis sessions are verbatim. And I was very much hoping that the process would dispell the whole notion of the visitors and prove that—despite appearances—the experience had been a complicated series of misperceptions.

Still, the study illustrated a very good point and revealed a fundamental difficulty even with serious and competent efforts to use hypnosis in dealing with this sort of material.

We just don't know enough about hypnosis to call it a completely trustworthy scientific tool in a situation like this. While Don Klein certainly didn't ask provocative questions, there is always the possibility that I was unconsciously eager to comply with an outcome that I might secretly have longed for. I might *want* powerful visitors to appear, to save a world that I'm pretty sure is in serious trouble. I'd spent the past three years working on books about nuclear war and environmental collapse. I knew full well that we are going to have a really rough time in the next fifty years. Maybe the idea of visitors coming along and saving our necks was more appealing to me than I might consciously have wished to admit. Maybe I hid my desperation from myself in order to live and raise a child with anything like a happy heart.

What I can say in favor of these transcripts is that they represent the response of an honest man to the efforts of a recognized expert in the field of medical hypnosis.

One of the greatest challenges to science in our age is from modern superstitions such as UFO cults and people who are beginning to take instruction from space brothers. Charlatans ranging from magicians to "psychic healers" have tried to gather money and power for themselves at the expense of science. And this is tragic. When one looks at the vast dollars that go each year to the astrology industry and thinks what that money would have done for us in the hands of astronomers and astrophysicists, it is possible to feel very frustrated. Had the astronomers been awash in these funds, perhaps they would have already solved the problem that I am grappling with now. I respect astrology in its context as an ancient human tradition. Still, I wish the astronomers could share royalties from the astrology books.

I did not believe in UFOs at all before this happened. And I would have laughed in the face of anybody who claimed contact. Period. I am not a candidate for conversion to any new religion that involves belief in benevolent space brothers, or in unidentified flying objects as the craft of intergalactic saints—or sinners.

And yet my experience happened to me, and much of it is recorded not in an unconscious context but in ordinary memory. If we are dealing with a new system of beliefs on its way to becoming fixed into religious dogma, the way the religion is in my case emerging, right into the middle of a mind with no obvious allegiance to it at all, suggests that real belief could be a totally misunderstood *biological* process

57

capable of occasionally issuing forth from some extraordinary and unsuspected structure of the brain far more concrete than Jung's collective unconscious. Thus, even if visitor experiences are an essentially mental phenomenon, to laugh at them or dismiss them as some known form of abnormal behavior when they obviously are not is in effect to be silent before the presence of the new. Science should bring its best efforts to this, which means good studies that proceed from open and skillfully drawn hypotheses.

If mine is a real experience of visitors, it is among the deepest and most extensive as yet recorded, and I hope it will be of value if they emerge. If it is an experience of something else, then I warn you: This "something else" is a power within us, maybe some central power of the soul, and we had best try to understand it before it overcomes objective efforts to control it.

What follows here are two transcripts of hypnotic regressions, covering my buried memories of October 4 and December 26, 1985. That these are buried memories and not imaginations worked out in the doctor's office seems hard to dispute. The mechanism that buried them is no different from that which places any particularly terrifying experience behind a wall of amnesia. Beginning with Freud, the process of screen memory has been extensively documented.

The hypnosis used on me was not qualitatively different from that used on police witnesses. And the same caveats that apply to police cases apply to this case—those and no others. It should be remembered, though, that—even given my earnest effort—I am describing what I *perceived*, which may or may not have been what was actually there. We really do not have enough experience with our reaction to extreme strangeness to know how we alter such memories.

Donald Klein met me in his subdued gray office on East Seventy-ninth Street in Manhattan. He is a tall man with curly hair and a quiet demeanor. Two things were immediately apparent to me about him as a hypnotist. First, I sensed command; he was confident of his skills. Second, he was a thorough, careful man with a very acute mind.

I had never been hypnotized before, and I was apprehensive about it. As it turned out, my apprehension was for the wrong reasons. I was afraid of relinquishing control over myself, which seemed deeply disturbing. Control, as may be imagined, was a central issue in a life such as the one I had been leading.

I found, though, that I trusted Don Klein when he told me that even under hypnosis people cannot be readily compelled to say things they do not want to say. I would not be out of control, not really.

The process of becoming hypnotized was pleasant. I sat in a comfortable chair. Dr. Klein stood before me and asked me to look up at his finger, which was placed so that I had almost to roll my eyes into my head to see it. He moved it from side to side and suggested that I relax. No more than half a minute later, it seemed, I was unable to hold my eyes open. Then he began saying that my eyelids were getting heavy, and they did indeed get heavy. The next thing I knew, my eyes were closed.

At that point I felt relaxed and calm, but not asleep. I was aware of my surroundings. I could feel my face growing slack, and soon Dr. Klein began to say that my right hand was getting warm. It got warm, and then he progressed to my arm, and then my whole body. I was now sitting, totally comfortable, encased in warmth. I still felt as if I had a will of my own, a sensation that was never to leave me. In fact, the hypnotized subject does have a will of his own, very much so. But he is also open to suggestion.

After some preliminary questions, preparing me by asking me to recall my birthday and then Labor Day weekend, Dr. Klein proceeded to the afternoon of October 4. I wish to add that Budd Hopkins was present at both of these sessions, recording them. He was allowed to ask questions, but only at the end of each session, and it was understood that his questions would be few. They are identified with his name in the transcripts. All other questions were put by Dr. Klein.

Events of October 4, 1985

SESSION DATE: *March 1, 1986*

SUBJECT: *Whitley Strieber*
PSYCHIATRIST: *Donald Klein, MD*

[This is an actual transcript of my first hypnosis. Nothing has been left out. This is what happened when my memories were examined under hypnotic regression.

Dr. Klein began the session with Labor Day. As I grew more comfortable with the process of remembering, he drew me closer to the night in question.]

"Now, we're going forward a little further, to the beginning of the month of October. Right around October first, 1985. Can you tell me where you are right now?"

"Yeah, I'm working on the Russian book."

"What book?"

"The Russian book."

"What's that?"

"It's a novel about Russia. I've got a good idea. I'm working on the Russian book."

"Where are you?"

"I'm at home in the city."

"You have any plans for the weekend?"

"Yeah, we're gonna take Jacques and Annie up to the country and I don't know whether or not Jacques is going to fit in the Jeep."

"Who are Jacques and Annie?"

"Jacques is a friend. Annie is his girlfriend."

"Now you're driving up to the country."

"Yeah."

"In a Jeep?"

"Yeah, it's a Jeep Wagoneer. We're not having any problems. Annie's very small, so Jacques can fit. He's in the backseat and my son's happy because he likes Jacques a lot. I put on a tape but nobody liked it. So we talked. 'I'm gonna take you all out to dinner tonight. It's too late to stop for groceries. We're gonna go to the—you want to go to the Top of the Falls?' We had a lot of trouble deciding about that. I remember that, but then we went to the Top of the Falls."

"Go forward to that time now "

"Yeah."

"How are you feeling?"

"Oh, I'm enjoying myself thoroughly."

"How far is the restaurant from your home?"

"Oh, not long, about fifteen minutes. And we had dinner. We had dinner. Anne, our son, all have a—I have a great time. Jacques has a good time."

"So you're back in your house."

"Yeah."

"And you're going up to bed for the night?"

"Yeah. I'm wearing my house shoes. We were all gonna sit in the hot tub, but I'm too tired." (I visualized myself in bed, talking to my wife before going to sleep.) "God, I wish I hadn't spent all that money on that restaurant."

"What happens after you go to bed?"

"Well . . . [Long pause.] Oh . . . I woke up in the middle of the night. . . . I don't understand that. Uh, there's something went past the window?" (I referred to an octagonal window beneath the peak of the living room's cathedral ceiling, approximately thirty feet from the ground. It can be seen from our bed and it looks out into a woods.) "What the hell? Something went past the *window? Something went past the window!* There's nothin' pa— Oh, God. 'Anne, the house is—' Something—"

"Something went past the window?"

"A big thing. [Beginning to cry.] No, it was a light! [Calmer.] It didn't go past the window. It couldn't have gone past the window. I'm going back to sleep. I think the stove's OK. It was a light in the front yard. I keep thinkin'. . . . Who the hell is that?" (I was looking into the far corner of the bedroom, where I saw a dark shape about three feet tall standing in the shadows.) "Is that somebody? [Pause.] Is that somebody *there*? That can't be. Y'know I'm lookin' at this thing. I don't think, I don't think I like that. [Long pause. Eyes open.]"

"Relax. Close your eyes. Relax. You're going to stay relaxed, you're not going to open your eyes. Stay relaxed. Now tell us what you saw."

"I saw something that looked like it had a hood on it, standing over by the wall near the corner in our bedroom [breaks into panic] and I don't want it to be there! I don't want it to be there! Please! God, it— What's it doing to me? Stop! Oh, oh, stop! What's it doing to me? [Screams, prolonged, twenty seconds.]" (I cannot recall experiencing at any time in my life such panic as was evoked at this point in hypnosis. My memory was of seeing the shape sweeping across the room and realizing with a feeling that galvanized my whole being that it was something totally unknown to me, glaring at me from right beside my bed in the dead of the night. I then emerged spontaneously from hypnosis. No written words, nothing, can convey my feelings at that moment. All I can say is that I relived fear so raw, profound, and large that I would not have thought it possible that such an emotion could exist.)

"Oh, I feel like I'm gonna throw up in a minute. I'm sorry. Oh, God. You know, I didn't know there was anything in my house till just this second that night on the fourth of October. [Weeps.] Ah, Jesus. Oh, God. Oh, boy. Scared the devil out of me. Sorry about that. I didn't expect this to be that bad, because if— I was prepared for being scared on the twenty-sixth. I did not know that anything had come into my house. . . . Oh, well. . . . It was there."

"Can you tell us?"

"You know it's dark. It's like a little man with a hood on or something. It looks almost like this—y'know, there's no head . . . he's covered in something. And comes over to the bed and he starts like sticking something in—not into my head, y'understand, but like it was sticking into my mind. It would make a noise like a voice. It was terrible!

[Demonstrates noise: a smacking, squeaking sound.] Like that. Going into me. It was just God-awful horrible. He was standin' there doin' that."

"What was he doing?"

"You know, I can't. I'd have to be hypnotized again if I was ever gonna find out what that thing was saying. It was something that was being said inside me. Like it had a little thing it could touch to my head and it would make a voice. That's what I think it was."

"While he was talking to you, you weren't shouting then?"

"I was shouting. No, I was." (I was confused by this question, because, as will be seen in a moment, the figure was not in fact talking to me.) "I have absolutely no idea why the other people had—didn't remember because I'm damned sure I was shouting."

"Do you remember what your wife was doing?"

"I don't see her, because I'm turned toward the thing. Am I still hypnotized?"

"No."

"I swear to God, I just can't believe that this happened. But it did happen. I'll tell you the light comes down past the window, then I see a glow in the front yard. I thought I had gotten up, but I don't think now that I did. I thought I had seen the glow against the roof of the living room, but I don't think I did. I think I knew all along it was coming in the window, the glow, and I just didn't somehow want to say that. Because it was very obvious even then that it wasn't a fire. And he was—I just don't know what he was doing. And I'll be frank with you all, I can't uh, I just can't figure it out. It was very scary. But he wasn't . . . He was wearing a cover, like, he had a cover on. I'm shaky."

"Perfectly understandable."

"You know, I can't understand it. Yes, I guess of course I understand why an experience like that could be so scary. But, you want to keep on?"

"If you'd like to."

"I do. Definitely."

"OK."

"I mean, I think I may have gotten right at the beginning through the worst part of it, frankly. Because whatever it was, when it first came to me that night it was ultimately terrifying."

Budd Hopkins responded. "It's often like this. Beginning moments are the worst. And after that, it gets easier."

"Was I a good hypnotic subject?"

"You were excellent."

"Good. It didn't seem to take very long. It felt very nice."

"It always feels nice."

"I was amazed at you because I thought you were explaining to me how it was going to be done, and the next thing I knew I was thinking I can't keep my eyes open."

"It's very simple, because once you get the hang of it you can do it just like that. Like this. Now look up at my finger." (In a few more moments, I was hypnotized.) "If you see something that is very frightening you will remain asleep. You will remain asleep, but you will tell us how you feel. The night of October fourth. You have woken up. You have woken up now. It's light. There is something in the room with you."

"It's dark."

"Tell me what you see?"

"When he sees I see him he comes over to the bed. He looks *mean*. He's little. Goes up to about the top of the lamp. Looking down at me. Got eyes. Big eyes. Big slanted eyes. A bald head. He's looking down at me. He's got a ruler in his hand. Has a tip of silver. Touches me. I see pictures. [Long pause.] I see pictures of the world just blowing up. I see pictures of the whole place just blowing up when he touches my head with this thing. [Weeps.] Jesus. It's a picture of like a whole big blast, and there's a dark red fire in the middle of it and there's white smoke all around it.

"Remembered voice: 'That's your home. That's your home. You know why this will happen.'

"'I know why. [Weeps.] Why don't you like me? Why do you hate me?'"

Dr. Klein: "Who said that?"

"I said that. [Pause.] 'Why'd you just put that thing on me, and it has the whole world blowing up? That's what I want to know. What is this *about*?! What is it about?!' " (Felt a sharp internal question, wordless.) " 'I don't know what it's about! When is this gonna blow up? What's gonna blow up?' " (A flashing picture of my son.) "I know what's gonna blow up. I know what's gonna blow up. I know it is, too." (I was then touched again.) "Oh . . . green. Shows me a park. I see my son. What's this got to do with him? Is this the devil? What the hell is this?

"Remembered voice: 'I won't hurt you.'

" 'I know you won't. I know you won't hurt me. Stop! Ah!' The house is burning down! The house is on fire. No it's not. That sounds stupid. Why did I say that?"

"Something woke you up. What happened?"

(I then emerged spontaneously from hypnosis.) "Explosion. I knew that. I was expecting it. I knew just when it would come. He took a little thing like a needle and struck it like a match in front of my face and it made a big bang, *pow!* And I thought the house was on fire."

Budd Hopkins: "Was he there after this?"

"I don't know. I don't remember a thing. When I said the house was on fire I'm already out of the bed. Right away."

"This was the explosion that Anne and your son—"

"I think the explosion that they all heard. Yeah. He did it with a little needle. Stuck it in the air."

"Why did he show you those images?"

"I don't know why he showed me those images, to be frank with you. They are the most dreadful images."

"Warday?" (A reference to my novel of the same name.)

"I'll tell you the truth, what I feel. I think he showed me images of the future of our world, is what he showed me."

"What about the green?"

"It's a beautiful, green expanse. Was immediately relaxing when I saw that. And my boy. My boy is in the park. My boy is there. And he's happy. That's what I saw. But—"

"Why are you so upset?"

"Because I think the park represents death, and he's there because he's dead. That's what I think."

"Why should the park represent death?"

"I don't know. That's just my impression."

"And the other scene of the world blowing up?"

"It's funny. It's not like of the world blowing up. I've gotta calm down here. I'm told it's the world blowing up. It's a red fire, a big, red, fierce fire with all these horns of smoke shooting out from it in every direction. And they said that's the world blowing up. I mean, Christ. I think we've got a monkey on our back."

Budd Hopkins: "No doubt about that. Other people have been shown this kind of an image, too."

"You know, I feel a tremendous relief right now. This is good to be able to remember. It's not an easy memory. But it's good to remember. Because I've been fighting to keep it out of my head."

"I can understand that."

"You know, I had a dream back in November. Of Cleveland blowing up. It was me remembering this image of the explosion but I thought it was a single city, so it wasn't so scary. Trying not to be scared of that image."

"You said he had a bald head?"

"He had a bald head and I would say he had slanted eyes. It was real hard to see because he kept putting this thing on my head. Almost every time I would move he would put this thing on my head."

"And then you would see an image?"

"I would see these images, yeah. I thought at first the thing was talking but when you hypnotized me again I could see the images. He had a little ruler thing that had a silver tip on it, quite silver, because I could see glimmers on it and it's dark in the room. Did you try to light up the room for me or something when you said the room was light?"

"I didn't."

"Because the room was dark. It was very dark."

"I thought you said it was light."

"Oh. I also have the impression that he was wearing covering. Like he didn't—when I looked—the thing that's scary in a way, the thing that scared me at first was realizing he had been there for some time, standing over there in the corner, and y'know I have a feeling about a fierce whoosh of some kind. . . . I'm not saying that I was being threatened so much as warned. That was my feeling. There was a very stern warning . . . or maybe it wasn't. Maybe if I had not been afraid of nuclear war and perfectly happy, when he touched that thing to my head other images would have come out. See what I mean?"

Budd Hopkins: "It certainly intersects with—"

"My own fears. Exactly."

"You spent a huge amount of time working on them in *Warday*."

"Maybe it was a guy making a psychological test of me. Could it be that? I mean, maybe he was literally testing me, doing something like, as a hypnotist, you might do, using that little silver object in place of your finger. Could that be?"

"It could be, among other things."

"Including that this was some kind of hallucination, but I don't think so."

"Where did he touch you?"

"Right here." (I touched the center of my forehead, just above the bridge of my nose.) "Very specifically right here. And every time he touched me there would be a burst of images."

Budd Hopkins: "You had no time to think between images?"

"No. This all happened quite quickly."

"They were visual?"

"Yeah, they were visions."

"Not words?"

"Absolutely not. They were pictures." (I became silent. I was aware of a great confusion of pictures in my mind.)

"Whit?"

"Yeah."

"You described two. Were there others?"

"Yeah, but the others are so jumbled up I can't tell what they were. You know, I think increasingly that they might be pictures out of my mind of my worst fears. Like nuclear war and my son being killed. And there's something else in there too that's all jumbled up. Maybe a fear so terrible that I can't even make heads or tails of it under hypnosis."

Budd Hopkins: "You said first that the figure seemed to be covered up, like with a hood."

"Yeah, but when it came close to me I could see its face."

"You said it had a bald head."

"Yeah? Did I?"

"Yes."

"Well, you see, I can sort of see that it had a bald, rather largish head for someone that size. And that its eyes are slanted, more than an Oriental's eyes. And they're *quite*— There's a piercing glare, almost. There's a real fierce look to the whole face. I'm not sure, but at some point I almost thought it looks like a bug. But not—you know, more like a person than a bug . . . but there were buglike qualities to it. Am I getting myself clear at all?"

"Oh, yes. Have you ever seen an image like that before?"

"I don't know. The only thing I ever remember reading about this was in *Look* magazine years and years ago. 'The Incident' . . . the John Fuller article about people who were picked up. That's all I've read about it. And whether or not they had pictures drawn I just don't know '

"Are you sure you haven't seen an image like that before. What about the book you have?"

"No, it has no pictures like that. I don't think so."

Budd Hopkins: "What about the Hynek book? I think there's a drawing in that." (He referred to a famous book on UFOs by Dr. J. Allen Hynek, *The UFO Experience*, which he thought I might have read.)

"I haven't read it. But you know, in our culture, there's so much media around. . . . It's possible, but I don't think so, because this is so damn real. It just seems impossible that it could be an image I picked up from somewhere—"

"It could still be real, and be an image that you—"

"Maybe the drawings were right. That's possible too."

"That's possible. It's also possible that it's something quite inexplicable that you're trying to hang something on, to give it some shape or form."

"Yeah, that's possible too."

"What I'd like to do is go back to that scene again, and see if we can't get some clarification about these other images."

"OK."

"I understand that this is a wearing business, and if you want to check out at this point—"

"No, I don't want to check out at any point. I'm determined to go through with this. And it seems now that there should be more, because it's obvious that this happened to me. And I would be highly irresponsible to myself not to continue."

"The pacing of it could be too fast. I don't want you to feel you have to be a hero in this thing."

"It's not a question of being a hero. It's much more a question of not wanting to walk out before the end of the movie." (Hypnosis was undertaken, using the same method as before, ending with a count up to ten. In about a minute I was rehypnotized.)

"Now I want you to go back to the point where you are having these images. And it's going to be like slow motion now. Everything's going to be going very slow, very slow, very slow. And very clear. Very clear. Tell me what you see."

"The world turns into a whole red ball of fire. It just seems to burst into flames like a little ball of gasoline out in the middle of the sky. And all these . . . smoke . . . things start shooting off it . . . like great horns made of smoke. And we're all there, down there in the red fire, in the middle of it. Then I see that thing on my head and it's gone. Picked it up off my head. Now I'm scared of him again.

Now I see . . . a park. . . . My little boy is sitting there on the grass . . . he's all wobbly, and he's like he can't move his arms right. He's all wobbly and his eyes look funny." (They appeared entirely black, without any whites at all.) "I have to go over and pick him up and help him. If I don't help him, he's gonna die. [Long pause.]"

(At this point there followed upsetting images of my father's death, images that did not reflect what really happened, but rather my fears about what might have happened.)

"And he puts that thing down on my head again. 'I miss you, Daddy. Oh, God, Daddy, why did you die? [Gasps.] Daddy, why . . . why—I just never got to know you, Dad.' Oh, God, my poor dad, died a hard death. Oh, she couldn't help him. It's my dad dying and my mother's sitting there staring at him like he was a little animal. Why couldn't she at least give him a good-bye kiss or something? I never knew it was like that." (I saw a clear image of my father lying on the couch in our old den, his head thrown back, gasping and choking. My mother was beside him in a chair, watching, too afraid to move. This was totally different from the scene she described, which was what would have been expected from the gentle and loving relationship that had emerged as his life came to its close. The image, though, was deeply shocking to me, and so real that I felt as if I could step into the scene. I then emerged spontaneously from hypnosis once again. It is very unusual to do this, especially from a deep trance like the one Dr. Klein had induced. It was an indication of the extreme severity of the emotions I was reliving.)

"Did that make sense?"

"Did it make sense to you?"

"Yes, it damn well did. It's a picture of my dad, lying on a couch going like that—gasping—jerking . . . and my mother's sitting in a chair, watching. And he dies."

"Did it actually happen?"

"I don't know. It's not the story she told. Maybe it's something I fear might have happened."

"Was your mother uncaring about your father?"

"No. They had their ups and downs in their marriage, but they were married for nearly fifty years, and I didn't think she was uncaring about him at the end."

Budd Hopkins: "So you feel these thoughts were maybe your thoughts?"

"They were my thoughts. They were definitely my thoughts. I mean, it sure as hell wasn't *his* father. He's pulling this out of my head is what—he's pulling it out of my mind. He's pulling things like my fear—perhaps there's a suspicion. First of all, when I saw that picture I felt an agony, because I never felt I got close to my father. My dad was distant. He was a loving father, but he always held something back, you know. He was from a very reticent generation. Rural Texans were very inward people. I guess I feel a little bit of guilt about that, or something. You know, I don't know what to make of all this. Do you suppose? I just don't know what to make of it."

"I don't really know what to make of it either, but it certainly sounds as if—"

"It's just—"

"You were opened—"

"It's so unexpected. This is the last thing I would have thought would have come out of me. And what's weird about it is, why would someone come from a flying saucer and evoke that kind of impression in me? What possible reason would they have?"

Budd Hopkins: "Well, that's not to find out now. That's speculation down the road."

"Like they were trying to find out how I ticked. It really is like that. Unless it's simply that I've come to a time in my life where there's some very difficult and terrifying material that I've got to face and this is how I'm facing it, and there was no little man there. But you know, I say that and I'm telling you right now that it's not true. It's not true. It's incredible, but it's not true. The man was there. He was standing beside my bed as real as life."

"You said, originally, about the December twenty-sixth

episode, that it was as if they said to you, when you disappeared, when your ego dissolved, if they had asked you what is your deepest secret, you would have told them right away."

"Right away. Yes."

"So you had an inkling about something. That your deepest secrets were coming out."

"Well, obviously I knew, because the memories were intact and they just came out of my head. Boy, though, if you'd asked me consciously I would have told you I had absolutely no idea what happened during that hour. If it was an hour. I thought I fell asleep after I saw the first glow."

"And the explosion seemed to come right after?"

"He took a little thing like a stick—a needle—and when he moved it even slightly in the air I could see it spark at the end, and he went like that [makes striking motion] and it went *bang,* and spread a tingling all over my face."

Budd Hopkins: "When I asked Annie Gottlieb what she would have to do to make the sound—I said, 'Suppose you were given resources to make the sound.' She said, 'If you had a big, heavy door and you pushed it back against the wall, *bang,* like that—'"

"It was a big noise."

Budd Hopkins: "And your Anne said it was like something hitting something. Almost like an explosion."

"Yeah. Well for me it was more like a—I can't say it was like a balloon popping, because that's too innocuous a sound. It had a heavier quality to it than that, like some big energy had been released."

Budd Hopkins: "Annie Gottlieb said it had a slapping sound—"

"Not that crisp. More of a thud. It was a big noise. There was a slap, but there was a deeper resonance to it."

Budd Hopkins: "That's what Annie said. We've got four different people to come up with a description of the sound."

"Thunder?"

"It wasn't like thunder, no. Not like thunder."

"It had a clap in it?"

"Well, no, because it didn't last after it. It would be like a clap that ended immediately. It had a deep undertone to it. But mostly not. Actually, a clap would be the best—like a deep clap of thunder that had no echo. Just a single noise. But it had a deep undertone to it. It had a very electrical quality to it. If you could make a tiny bolt of lightning in someone's face, you would create thunder right in their face. That's what was done."

"I think we're about finished."

"Yeah, I don't want to go into the twenty-sixth now!"

"We might not be finished with the fourth."

"Now that the fear is over. The turmoil. I feel I don't have a psychiatric disorder. I feel you're right about that. You know what I've got to do? I've got to figure out how I feel about this, because I don't think I'm intellectually going to be able to deny their existence much longer. And I have to understand how to feel about these beings who would come into my house and do something so strange and yet somehow or another so productive."

"How productive?"

"Well, in two ways. One is, they learned a lot about me, if they are interested in me, for whatever reason. This afternoon I just learned a lot about myself. I learned a lot. Things I didn't have any idea worried me. About my dad and mother."

"The other fears—"

"Well, fear of war, obviously . . . and of the death of my son. There is no such thing as a good father who doesn't worry about harm coming to his kid. But the other material is a great surprise. And that's as vivid as it can be. I loved my mom and dad so much. I love my mom, still, and I want to believe that at the end it was as gentle and loving a moment as Momma has always said."

"That isn't an image of what really happened?"

"No, I have no reason to believe that. Maybe something much more subtle is going on here. Maybe that image was created to see how I react to something that would be ultimately terrifying to me. Or maybe they were just trying to find out what kind of person I am."

The session then ended with a decision to continue later in the week. The next night (Sunday, March 2) I called my mother in San Antonio, as I try to do every week or two. I told her nothing about this matter. And how could I? I had not thought of a way to explain what was happening to us to my seventy-year-old mother on the telephone.

We talked for a time about a friend who was in the hospital. Then, without warning, she suddenly described my father's death to me. I did not ask her to, nor was I even hoping that she would. In the past ten years I have heard this description only once before, the day after he died. She recounted how she had been sitting near him while he lay on the couch. He had spent a restless afternoon. The doctor believed that his heart would soon fail, and had told my mother this just a few days before. Still, they had been together for so long she could not imagine him dying.

In the last years of their marriage they had become extremely close, often sitting hand in hand together, in the wordless communion that sometimes blesses very old relationships. I can hardly imagine a more gentle or loving end to their long time together than what happened at the last.

Mother told me again how she had suddenly heard Dad call her name, and had gone to him and said, "Karl? Karl, wake up." He was lying still and silent. . . . It was as easy as that.

How was it that she would suddenly retell this story again, after all these years, at the very moment I needed to hear it? The combination of the memory of that terrifying night and this story, told in my mother's calm, sure voice, led me into the most enriching of insights about my buried fears and guilts. I blamed myself for the lack of intimacy in my relationship with my father. He reached out more than he withdrew. Even though I loved him, I moved away. I grew up and left him to age and die without the comfort of his oldest son.

Also, though, I had to make my own life. Beyond its moral sense, the word *conscience* has always meant to me an active knowledge of one's inner truth, an acceptance of all the sacrifice on the part of others that has been required for one's own development. The prime sacrifice is that of the parents. One can preserve the guilt one feels for it—as I now see that I had done—or one can temper it with acceptance and use it as a building block in the edifice of maturity. In a moment that night, beneath the feather-pounding of the silver wand, I was given a potential that could greatly enrich my life.

If this was a real visitor, giving me a real blessing from some other reality, then why was it hidden in amnesia where I could not gain access to it? Maybe my experiences were only a side effect of some sort of study. Or maybe it was known even then that this rich treasure would eventually be open to me, because the whole experience had been designed in detail by insightful minds engaged in a slow process of acclimatizing humanity to their presence.

Maybe, though, there was another truth here. Perhaps the hypnosis revealed not just the possible presence of visitors but the action of a hidden and tremendously therapeutic potential which, if correctly marshaled, could be of great value.

While there is a long tradition in the fairy literature of the Middle Ages of the use of wands to grant insight, and the angel in the Book of Revelation is said to strike the elect thrice between the eyes and cause them great suffering, modern accounts of visitors contain only one oblique reference to this process. A woman who had an enigmatic visitor encounter in the fifties slowly became insane thereafter. As she did so she would claw at the center of her forehead in the same place where I was struck with the wand, to the point that she gouged herself almost to the bone.

It would be easy to say that the material revealed here is the work of a mind making opportunistic use of some nocturnal disturbances to gain contact with fears that it needed to explore. The glaring difficulty with this supposition is that the whole transaction remained hidden in amnesia until many months later. There is the additional problem of the witnesses, and the "clap of thunder" coming before the "lightning."

The easy route would be to dismiss this material as entirely psychological. That would also be a mistake, at least until the physical effects are explained completely, in detail, and satisfactorily.

A terrifying thing happened to me. Perhaps it involved visitors from somewhere—maybe even from inside the human unconscious. For me, though, the most important thing about it was its essentially *human* effect. I was a human being, and my part of things involved having a human experience. Even if there was a visitor, it seemed clear that concentration on the human part of the encounter was the key to understanding what meaning it may have for me. And the visitor was no more than wind in the eaves or the moon lighting the fog . . . then it was a key to what I mean to myself.

Events of December 26, 1985

SESSION DATE: *March 5, 1986*

SUBJECT: *Whitley Strieber*
PSYCHIATRIST: *Donald Klein, MD*

We met again a few days later. I had occupied myself with other things during the previous four days, but it was hard. It was a great effort not to go to the library and get half a dozen books about close encounters, and another half dozen about possible psychosocial causes for such experiences. But I agreed with Dr. Klein and Budd Hopkins that I must remain as ignorant of this material as possible until after my hypnosis.

Yet I kept remembering that face, darting, the sharp

dark eyes glistening, and the silver wand glittering as it rose and fell.

I couldn't believe it could be anything other than an act of mind. While I was prepared to accept that there may be a visitor presence on earth, I was not prepared to find one of them at my bedside practicing psychotherapy with a fairy wand. Surely it wouldn't be that personal. Surely it would be at least *a little* like what we would expect.

But there are deep, deep waters running here. If these are indeed visitors, they know us well . . . better than we know ourselves. More than visitors, they may simply be "others," an aspect of being which we have not yet understood.

No matter what exactly is made of it, the combination of all the flying-disk sightings over the past two generations and the smattering of abduction accounts certainly suggest that something strange is going on. Maybe just a strange form of hysteria, but if so an awfully strange one . . . that combines huge lights, little scampering feet—and intimate intrustions into the soul.

Budd Hopkins told me that first hypnosis sessions were often traumatic. These memories are buried for a reason: They are frightful in the extreme. When they first emerge, the mind lives through the panic it has been avoiding. While my experience with the wand is almost unique, the being I saw wielding it is of a type commonly reported.

It was during this week that I began to have a relationship with my own memories. There had been a being present. I had seen it. And I had seen others in December. I remembered the way they had smelled, the way it had felt to be carried by them, the way their place had looked inside.

I felt complex emotions, ranging from the deepest inner unrest to what I can only describe as an urgency to compliance. I wanted to come together with them on my terms, to find some sort of mutuality.

I have never felt so tiny, so helpless. My boy's words haunted me— ". . . a bunch of little doctors who took me out on the porch . . ." There is nothing so hard as being a parent frightened in the night for your child.

When I returned to Dr. Klein's office, I described myself as "uneasy." He said, "Is that all?"

I admitted: "Terrified."

"Very understandable."

We began the session covering December 26 as soon as I was comfortable. Again, Budd Hopkins was present and allowed to question me under the same rules agreed to in the previous session.

"I want to take you back to December twenty-sixth. Going back to December twenty-sixth. And you are having supper. You are going to talk to me now, but stay completely asleep. Completely, deeply asleep. Where are you having supper?"

"In the country."

"Tell me who's there."

"Anne and our son."

"How are you feeling now?"

"Nice."

"What are you doing?"

"We're having supper."

"What are you eating?"

"Goose. Cold goose. It's a used supper . . . Christmas dinner. And cranberry sauce. Sweet potatoes."

"How do you feel?"

"I'm very happy. I'm feeling great."

"Had you been feeling great the previous few weeks?"

"[Long silence.] I had a hard time up until Christmas."

"What sort of hard time?

"[Long pause.] Was—scared. Unhappy. I felt like the world was caving in on me. Kept thinking there were these people hiding in the closet. Went all through the house every night. Checking."

"Were you checking anything out?"

"I was checking out the house."

"Did you have any idea why you were searching?"

"In case there might be somebody hiding in the house."

"Who might be hiding?"

"People. Them. Those people."

"Did you know about those people then?"

"Yeah."

"What did you know?"

"They might be hiding in the house."

"Did you tell anyone about it?"

"No. I didn't know about 'em."

"Did your wife ask you why you were looking in the closet?"

"No. She never saw."

"You hid that from her?"

"Yeah. And from my boy. I've got a gun."

"What sort of gun?"

"Riot gun."

"When did you get that?"

"October."

"When in October?"

"We went to the gun store. About . . . the leaves were falling. . . . I don't know, I think it was . . . in October . . . middle of October . . . but I went right out and got it. . . ."

"What did you want it for?"

"Protection."

"From what?"

"Not sure. I just have the feeling sometimes . . . there are people in the house."

"I'm going to take you forward in time, forward in time to the evening of the twenty-sixth. You're going up to bed to go to sleep."

"Yeah."

"Going up to bed to go to sleep now. I want you to tell me everything that happens. You are going to remain calm, and tell me what happens."

"We go to sleep in bed. I have a real good book to read. Give Anne kiss. Can hear the snow on the house, a little bit of snow. Turn out the light. I go to sleep. Did I turn on the burglar alarm? Hmm. I listen to the radio for a while. They have on 'Our Front Porch'? Jazz. S'late. Turn it off and the whole place fills with quiet. Now I'm getting sleepy. Go to sleep. [Long pause.] Definitely . . . think . . . I hear 'em. I hear them. Comes right in the door, looks like he's wearing—cards. God damn! I can really see this! He looks like he's wearing cards. On his chest, this big, square blue card on his chest. An oblong one down on his middle. And he has on a—a round hat. And he's wearing a face mask with two eyeholes and a round hole between them down toward the bottom and he's moving real fast. And he makes— stands beside my bed and makes a gesture to the door. And there's a hell of a lot of them! Filing into the room! I'm talking about a lot of them. They're not wearing the cards. They're wearing overalls—coveralls—blue. I can see their heads, which are bald. Time to get up. I get up. I'm scared, you see. I'm scared as hell. I take off my pajamas. Scariest to see Anne. I have to say good-bye to her now. There's two whole rows of 'em. I'm going out. They're moving me. They're moving me. They're moving me. Are they moving me from hand to hand? They're little bitty people. I feel like I could almost pick one of them up with one hand." (I was taken downstairs onto the front porch, where I saw a sort of black iron cot.)

"I don't like the look of that thing. That's a cot, or like a bed. Only—it's for me. I feel sick inside. Just *sick*! I'm not—I'm not. I'm just sort of watching all of this. Because it always starts like that where you don't know it's not a dream, but it's not, you know. And when they came in, like two big lines of them came right in the door. The whole room. I mean, there were a lot of them. You're talking about a whole bunch of them.

"I don't know where they are now. I lie down on the cot. It just gets . . . sort of jumbled. I remember I thought it was almost like getting into the electric chair. And it goes off the back of the porch and I know this is a dream because I'm flying. I'm flying so this must be a dream. I don't want to see the last of that house, though. I don't. I don't want to see the last of that house. [Sobs, gasps.]

"Where is the snow? We're way out in the woods, way the hell. You know, I can't understand this because I've gone so far from— It's like there's vines right on my shoulder. [Long pause.]" (I was trying to hear one of them, who was explaining something to me. But I could not repeat what I heard. Whatever it was, it terrified me. I opened my eyes.)

"What woke you up?"

"I don't know if I can tell you. And also I'm not sure I'm awake. Am I awake?" (The room had a vague look to it, and I felt deeply relaxed.)

"Hard to tell at this point. Why don't we try to go back in. Very quiet. Relaxed . . ."

"It was real clear at first. Then just this second it got—"

"Things will be slowed down for you. We'll go through this very slowly."

"OK."

"Slow and relaxed. Slowing down for you. Back to this time. Are you floating?"

"No, they're carrying me. At least they're around me. You see, what's so funny is I'm lying down and I can see the sky. I can see the clouds. And—they're all around me, though. And I'm naked and I'm not cold. And I can see the sky. This thing has like . . . there's two places for my arms . . . and it's not like a bed. It's got places for my feet and it's got a place for my head. I'm lying on it and I'm looking up at the sky and I can see . . . things—like the clouds. And they have a—there's like a swarm of them, they're around me. And I have a feeling—I don't feel very much. I feel very numb and funny. Not bad. It's kind of nice. It's a feeling like you're just sort of numb. And the next thing I know I'm sitting. I'm still in this thing but I'm sitting in the woods. It's almost as if it became a part of me. Like a— It keeps me in it, you know, but if I move around it stays with me. And I'm not going anywhere, that's for sure. Because this thing just stays right with me.

"We're sitting in a little—we're sitting in a hole. Like a little hole. I'm just like this. [Shows a semireclining position, hands as if grasping the arms of a chair.] I remembered someone sitting right over there, but now I don't remember it. Or they're not there anymore. [Pause.] These people are scaring me. Terribly, because . . . *whooof!* Right up! Just

shot right up. Yeah, that scared the dickens out of me. I saw
the trees down there! Now they've got a floor under me
again."

"You went up?"

"I just shot right up out of the woods. In this chair, this
thing. I just went *whoompf*, right up out of the woods."
(Other people have reported this sailing up into the air also.
I did not see what I entered.)

"Very high?"

"Yeah, you're telling me! I went way up. Right up. I
must have gone up a hundred feet. More than a hundred
feet. Up past the trees. Just like that. *Whooompf.* Right up
out of the woods. It was like going up in an elevator. Really
felt it. I mean, I felt it. Right up. *Whoof!* And now they've
got a floor under me. I know this is no dream. I sure hope I
get home again. I'm glad they took that thing away. I'm
sitting on a bench in a little room. [Sniffs.] And it smells
funny. Smells somethin' like cheese in here. Smells kind of
nasty, to tell you the truth. It's not clean in here. Here's
something. Somebody talkin' to me. There is somebody
talking to me. Now she walks right past me in the front.
And she's wearing a tan . . . suit. She looks like a little person made out of leather, sort of. I see the head real clearly,
and you got—you got—it like, you know, makes you sick,
kind of, because you know—I can't—I don't know where
this thing could be going." (The little room, usually round,
is almost a universal experience. A tan suit is also common,
as is the description of the skin. Gloving leather has been
described. This being, by the way, did not look even a little
human. Her looks coincide with one type of visitor that is
often described, especially the eyes.)

"'You know what, I think you are old. Are you old?'

"She says, 'Yes, I'm old.'

"She's lookin'—lookin' at me. [Moves head back, then
to left and right, as if being held by the chin and examined.]
She's lookin' real close. She's got a matchbox. No, it's not a
matchbox." (In this exchange, I remembered a deep, basso
profundo voice. She then told me that an operation would
be performed.) "'Aww, what is it? What do you mean, an

operation? What do you mean, an *operation*? What do you mean, an operation?' I'm getting real scared again. Real scared. Because I cannot do a thing about this. I don't even want to look up at this.

"'Can we help you stop screaming? Can we help you stop screaming?'

"'You could let me smell you.' She puts her cheek up by my face. They are here. You have to understand that. They are here. 'I'm not going to let you do an operation.'

"'We won't hurt you.'

"'I'm not gonna let you do an operation on me. You have absolutely no right.'

"'We do have a right.'

"That was it, bang. There was nothing to it. I thought they were gonna cut my whole head open. There was nothing to it."

Dr. Klein: "What happened?"

"Just a bang back behind my head, that's all. Not loud. Just bang. She's sittin' right in front of me the whole time, just lookin' at me. They're moving around back there." (I could sense them, but I was looking at her. She drew something up from below.) "'Jesus, is that your penis?' I thought it was a woman [Makes a deep, grunting sound.] That goes right in me. [Another grunt.] Punching it in me, punching it in me. I'm gonna throw up on them. [Pause.]" (They began trying to open my mouth with their hands.) "'What do you keep wanting to do that to my mouth for?' They keep trying to put something in my mouth. They're real. They're *real*. Put up her cheek right to me, and they're real! That's the incredible thing here. I've still got this thing in me and it'd be nice to take it out. [Pause. Long breath.] I had a chance to look around in here. And there's a bench, and there's something like a pair of old clothes lying over there on the side. And there's a door. A round door. And it's closed. It has a little black nubbin in it. In the middle of it. And it's closed." (I heard a murmur.) "What the hell did she say to me?

"Voice: 'You are our chosen one.'

"'I don't believe that for a minute. It's ridiculous.'"

83

(They asked how I knew that it's ridiculous.) "'How did I know that? Because it *is* ridiculous. Sing that song to somebody else. And also I want to go home.'"

Dr. Klein: "What did they say?"

"'You are our chosen one.' And it's bullshit and I know it right away. S'like a joke, almost. She says, 'Oh no, oh no.' [Imitates singsong.] Y'know, like they're trying to pull my leg. I want to go *home*.

"'What if we don't let you go home?'

"But I don't know if she said that or not. I think I think that she said it." (I was shown that door again, which for some reason terrified me. I was asked if I wanted them to open the door.) "'I do not want you to open that door! I belong with my momma and my wife . . . and my boy. That is where I belong. [Sobs.] I don't belong here. I don't know how I ended up here. What the hell did I do to attract all this?'" (She asked if I was as hard as I could get. I did not know exactly what was meant.) "'I guess I am.'

"Voice: 'Can you be harder?'

"'Can I be harder?' Oh, Lord. Didn't know I was hard like that. 'No, not with you around I can't be harder.'

"Voice: 'What would you like me to be?'

"'What would I like you to be? I'd like you to be a dream, is what I'd like you to be.'

"Voice: 'I can't be that.'

"'I know you can't be that. You just cut me on my finger.' Just like that. He just comes up, pow, bang, gone. Doesn't hurt at all. Doesn't hurt at all. I'm not scared the floor goes away like that. I'm like rolling like a ball. Feels like I'm going backwards in a movie, almost. It's like you just had total freedom and you could fly, only you're not going anywhere except down the rails. Oh, boy."

(I went sailing right back into my living room in no more than a minute. I had no memory of where I had just come from.) "I sit on the couch. I think I'm gonna build up the fire except I haven't got any clothes on. So cold. So tired. I go upstairs. There's two people standing up here now. And it scares me because I'm—I don't think they were there. I don't think they were there. I go in the bathroom, brush my teeth. I can't get that face out of my head. I sure am glad to get home. Now I have to just go to bed. I see my dark pajamas, blue pajamas, put on my pajamas, tie them up, button them up, get right into bed. I touch Anne, and she's warm. God, I wish I could live in a prison. [Sits with eyes closed, slumped as if sleeping.]"

"I want you to relax. I want you to go back, back. I want you to see her very vividly, very vividly. I want you to see her face."

"Yeah."

"Why do you say she's a woman?"

"I don't know. I just think it is. Old, too. She's got bald . . . she's got a big head and her eyes have bulges . . . she's sort of brown-skinned, not like a black person but like leather. Yellow-brown. And when she opens her mouth her lips are all—she hasn't got lips exactly—but it flops down. Her lips are floppy. I never saw her talking to me. You know, the truth is, I don't know what that is. I don't know whether it's a bug or what. And I also don't know if it's a woman or not. [Speaks in high, light voice.] It talks like this. It's got sort of a— [Normal speech, as if to creature.] 'You know, I'm not buying this. You can show me all that little insignia you want, and I'm still not buying this.'"

"What did she mean by saying can you get harder?"

"I was about half up. Hard. Penis. And she says, 'Can you get harder?' And the truth is, I could not. I didn't even know I was in that state. And with her around, there's just no way."

"Was this natural, or somehow induced?"

"I don't know. No. But you see, that thing stayed in me. I don't even know when it went out. It was almost like it was alive. It was a big, gray thing with what looked like a little cage on the end of it, a little round nubbin about the size of the end of your thumb. And they shoved it into me . . . they showed me afterward . . . so they must have taken it out of me, but I don't remember them doing it. These things happen sometimes like they're sort of in between. [Pause.] You know, they talk to me, but I can't hear 'em. [Long pause. Sigh.]"

"One thing you mentioned was this message that you were the chosen one."

"Yeah."

"Did you react to that?"

"Yeah. I said exactly what I said then. Because they say, 'You are our chosen one,' and it's just bullshit. Like they're trying to stroke me, you know."

"Did they say chosen for what?"

"Nah. Not for anything. They've got a lot of them, believe me. I've seen some of the others before. All lying down there."

"What others?"

"The other people. There was a whole row of 'em. But that was a long time ago. They didn't know where they were or what they were doing. I was sittin' up in bed. And uh—"

"How old were you?"

"Twelve."

"Where were you?"

"I was sittin' up in bed. And everybody else was asleep. There's a whole bunch of beds . . . [Sounds of distress. Long sigh, as if resigned.] 'I'm glad you let me be awake.' I'm sittin' on a chair . . . just this gray thing in front of me. 'What is *that*?' It's got red spots on it. 'I'm tired. I feel sick.'

"'Do you want to go home?'

"'I don't care if you never take me back home again.'

"'You have to go home.'

"'Who are all those people?'

"'They're all soldiers.'

"'Why'd they end up in here?'

"'Because they were alone.'

"'What do you do to them?'

"'We look them over and send them home.'

"'What's the point of that?'

"'The point of that is—the point of that is—*well.*'

"'Why do you look so awful?'

"'I can't help that.'

"'When did you find my sister?'

"'She's just down the hall.'

"'Patricia? E-p? E-p doesn't look good. E-p looks like she's deader'n a doornail.'

"'She's all right.'"

(I then saw my father for the first time. He was standing up, apparently quite conscious.) "'Daddy!' I'm scared now. They've— 'Daddy! Don't be so scared, Daddy! Dad, don't be so scared! Daddy, don't be so scared! Oh, Daddy! Daddy, don't be so scared! Come on, Daddy. Daddy, it's all right!'

"He says, 'Whitty, it's not all right! It's not all right!'

"'No, I know it's not all right.'

"'Oh God, what is it?' he asks.

"'I don't know what it is either, Daddy. How'd you get up here, Daddy?' [Gasps, stifled screaming. Slowly subsides. Long breaths. Silence. Emerges spontaneously from hypnosis.] We were on a train. Were we on a *train*? [Long pause.] I'm not hypnotized."

"Do you recall what went on?"

"Do I recall what went on?"

"The last part of your hypnosis."

"The last part? We were on a train. I was scared to death. Just scared to death. Something had happened to my father. Is that—is that true—is that what I? No, because it's not true. We weren't on a train."

Budd Hopkins: "You talked about your sister."

"Yeah, my sister was there with us."

"Edie?"

"E-p."

"Ebie?"

"No, E-p. Did I say E-p? That was her nickname."

"Do you call her that?"

"That was what we called her back when we were kids. How did I end up back in the— I'm a little confused, because, uh . . . I remember saying I'm twelve at some point, and then I remember seeing that thing again, the same thing that I saw when—"

Budd Hopkins: "What was the thing?"

"The thing is a—why, I keep saying it's a woman, you know. But it's a thing. But I saw her. On the train? What in the world is this all about, because I seem to remember seeing her on a train. *On a train?* But it's not—it's just not a train. I'm telling you that right now, Budd."

Budd Hopkins: "Your father was there?"

"Yeah. He was there. He was scared to death! And when he got scared I got scared. And my sister was there but she was out like a light. And there was a whole bunch of soldiers there too."

"Regular soldiers?"

"In uniforms."

Budd Hopkins: "And they were out, too?"

"And they had uniforms. They were all lying on—"

Budd Hopkins: "Unconscious?"

"Tables—no, they were beds, but they were solid—no legs. They were going out in both directions, sort of."

Budd Hopkins: "Many?"

"Lot's of them. Yeah."

"You were allowed to sit up?"

"I was sitting up. I was happy and sitting up. Very excited. Then the next thing I knew I saw my father and I was terrified because he was so scared."

Budd Hopkins: "Was this the same scene or a different place?"

"No, this was the same place. It was not quite. I was sitting in a chair and there was a gray thing in front of me like a gray box that came down—totally gray—I could see the edges of it and the bottom but not the top. Because I was restricted sort of in my movements. You know, I keep thinking that this was on a train. I'm still thinking that, but it can't be, can it? I'm talking about memories that didn't happen on a train obviously, am I not? Did I say it happened—you know, I'm beginning to—I'm very confused here! [Laughs.] I don't know what the hell's going on."

Budd Hopkins: "What made you say *train*?"

"I don't know."

Budd Hopkins: "Did you see—"

"No, no we *were* on a train. We really were on a train! We were on a train, and I'll tell you, the goddamndest thing happened when we were on the train. We ended up in this thing when we were on a train. The three of us were on a train. I'll tell you what we were doing, too. We were coming back from Madison, Wisconsin, on a train. In the year 1957, and that's when all this happened. I have no idea how I ended up there from being on a train."

Budd Hopkins: "Did you see like seats going back—"

"No, no, we were—you're kidding. My father didn't go on trains in seats. We were in a great big drawing room."

Budd Hopkins: "You were in a drawing room, OK."

"Yeah. Together in a room. And all of a sudden I'm not on a train, I'm sitting up in bed, and all these soldiers—"

"Did you have any covers?"

"No, no—yeah—there was a little—don't go so fast for me, Budd. I know you're eager to know, but my mind, my mind keeps—there's something in me that keeps saying, 'You're on a train, you're on a train, you're on a train.' And it's like I'm—it's very hard—but no, there weren't covers. There was—my impression was there was something soft under me. It wasn't an unpleasant place to be, in that sense. It had solid sides that came up a little bit above the edge. Then I was sitting up in it. Then I was sitting on the edge. You see, I don't remember moving. That's the thing that's funny about it. I remember being in one place and then in another place. That's the damndest thing. I never remembered anything like that before."

"Now in that situation, when you're sitting up, was your father there?"

"No, I sat in front of this gray thing for a while, in a little chair. And then, all of a sudden I saw my sister down here, kind of [points down and to the right], lying there just totally out. And I was real surprised and scared, and I feel scared again. Then when I saw my father he was standing up and he looked totally bereft and terrified. Scared, so scared. And he put his head down and started doing this, and it just scared the hell out of me." (I made a convulsive mouth movement, imitating my father. It was as if he was trying to get something out of his throat.) "And then I heard him screaming, but real faint, you know. I could see him—he was no farther away than you are" (about four feet) "but I could hear him very faintly, just screaming and screaming. The second I heard him it put just a—a—terrifying fear in me. I remember that, right sitting here, the way that felt, it just went right through me. It was worse than the last time, last week, only the difference is that when you started out you said to be calm, and it was breaking through that, and that's why I woke up. When Daddy was scared, I was just scared to death."

"You said you took a trip."

"Yeah, the trip happened. And not only did the trip happen, something did happen on the trip. Because, on that trip, on the way back I was as sick as a dog. Vomiting and vomiting up bile. And my father was just having a hell of a time. God, he must have had a rotten trip, poor man."

"You mean—"

"He was having a hell of a time with me because I was so sick."

"Your sister still alive?"

"Yeah."

"There was something—you were describing as if you'd seen it before."

"You know what I saw before? The woman. The same person."

"That occurs in hypnosis. We have these spontaneous age shifts. Frequently what will happen is that somebody will under hypnosis see something they've seen before, or something like it, and age shift will occur."

"I remembered vaguely before hypnosis that I knew someone there. But I just put that out of my mind because that's impossible. You can't—I mean, it's one thing to deal with something like that, and an entirely different thing to find out you know one of them already. [Laughs.]"

Budd Hopkins: "What about this thing about the woman—"

"This is just so strange! Will you stop for a minute, Budd, I just can't stand this. I mean, it's just—we're gonna have to talk about this another time because I just need to rest."

"Let's go up and relax."

"Yeah, I've just had enough."

THREE

Farewell, green fields and happy grove,
Where flocks have ta'en delight;
Where lambs have nibbled, silent moves
bright:
The feet of angels bright;
 Unseen they pour blessing,
 And joy without ceasing,
 On each bud and blossom,
 And each sleeping bosom.
  —WILLIAM BLAKE, "Night"

THE COLOR OF THE DARK

Insight

Lost

When I left that last session, it was with a deep sense of concern. I felt that I was entering an unknown region of the mind, perhaps of experience. I was doubly worried now for my sanity. First, I still felt that I might be the victim of some rare disorder. Second, I questioned my ability to live with the notion that my whole life might have proceeded according to a hidden agenda. Neither of these alternatives was acceptable—hardly endurable—and yet one of them had to be true.

I walked the streets of New York, not thinking, just absorbing the comfort of ordinary life. I walked, but my impulse was to run. I was trapped: If I did not accept that something real was hiding in the deep of my life, then I had to accept myself as a disturbed man. But I did not feel or act disturbed. I felt afraid, and all my irrational actions could be seen as a response to unacknowledged fear.

I was a responsible husband and father. There wasn't any sign of psychosis in my personality. Don Klein was an acknowledged expert and, even after this hypnosis session, he told me that he thought I was sane.

But how could anybody not be psychotic and yet have these spectacular delusions?

As I walked I considered the problem. What should I tell my wife? And how about my son? To what degree were they involved? Never in my life had I felt as I felt then: trapped in a mysterious cavern of a life that had once seemed so clear and understandable.

How could it be that this went back into my childhood? How could it *be*? And if that wasn't true, and my mind had chosen to do this to itself, then what was it doing, and why?

Much later I listened to the tapes of other people's memories and hypnosis sessions (with their permission) and read Budd Hopkins's book, *Missing Time*. I then participated in a colloquy with other people who remember being taken. There I found that multiple-episode memories are quite commonplace. Many people who report being taken report a lifetime pattern much like the one I had discovered.

I wrestled with the notion that something might have been happening in my life—real encounters—that were having a tremendous, hitherto unconscious effect on me. Certainly I had acted as if this were true before any conscious memories had emerged. The conscious memories didn't really come before the first week in January 1986. Yet, as early as the summer of 1985 I had become nervous about "people in the house," even to the point of buying expensive burglar alarms and, in October, a shotgun.

I even tried to move—back to central Texas, where I grew up. It is interesting to note that I was, if anything, even more fearful in Texas than I was in New York. Did this mean that I unconsciously recalled even more frightening things happening there?

When I finally got home to my apartment it was to a bright, warm household with dinner waiting. Ten minutes later I really felt as if I had left the shadows behind.

But then my son went to bed, and soon after Anne turned in. When the lights were low my home seemed no more sheltering than a place of air.

When it was time to be alone in the night, what I now had to take with me was a corpus of staring owllike faces, a shockingly revised personal history, and a great deal of fear.

That night I wished to God that I could somehow shed myself and step out fresh in the world. The visitors persisted in my mind like glowing coals. I could see those limitless, eternal eyes glaring right into the center of me. Visitors seemed to inhabit every shadow, to move in the course of the sky.

I went out again and walked some more, going down through SoHo and into the empty streets of TriBeCa.

When I finally went back to the apartment the cats came

up to me and started to twist and turn around my ankles—and then went bounding away. My cats. I shut myself in my office and sat cross-legged on the floor, trying to collect myself.

As soon as I relaxed, it was as if I had opened a hatch into another world. They swarmed at me, climbing up out of my unconscious, grasping at me. This was not memory, it only worked through the medium of memory. It was meeting me on every level, caressing me as well as capturing me. This emergence was like a kind of internal birth, but what was being born was no bubbling infant. What came out into my conscious mind was a living, aware force. And I had a relationship with it—not a fluttering new one, but something rich and mature that ranged across the whole scale of emotions and included all of my time. I had to face it: Whatever this was, it had been involved with me for years. I really squirmed.

What might be hidden in the dark part of my mind? I thought then that I was dancing on the thinnest edge of my soul. Below me were vast spaces, totally unknown. Not psychiatry, not religion, not biology could penetrate that depth. None of them had any real idea of what lives within. They only knew what little it had chosen to reveal of itself.

Were human beings what we seemed to be? Or did we have another purpose in another world? Perhaps our life here on earth was a mere drift of shadow, incidental to our real truth. Maybe this was quite literally a stage, and we were blind actors.

To gain some semblance of control over myself, I decided to make an inventory of possibilities. I sat down at my desk and began to write.

Even if the visitors were real, there was no reason to believe that they were simply creatures from another planet.

I speculated. It could be that the "visitors" were really from here. Certainly the long tradition of fairy lore suggested that something had been with us for far more than the forty or fifty years since the phenomenon took on its present appearance. The only trouble with this theory was that what has been happening since the mid-forties seemed more than just a little different from the fairy lore. Now there were brain probes and flying disks involved, abductions and gray creatures with staring eyes. Surely no change had taken place in the human psyche extreme enough to account for such a radical change in the appearance of the fairy. And yet, there was undoubtedly something here. . . . I thought perhaps the visitors were somehow trying to hide themselves in our folklore.

Another thought was that the visitors might really be our own dead. Maybe we were a larval form, and the adults of our species were as incomprehensible to us, as totally unimaginable, as the butterfly must be to the caterpillar. Perhaps the dead had been having their own technological revolution, and were learning to break through the limits of their bourne.

Or perhaps something very real had emerged from our own unconscious mind, taking actual, physical form and coming forth to haunt us. Maybe belief creates its own reality. It could be that the gods of the past were strong because the belief of their followers actually *did* give them life, and maybe that was happening again. We were creating drab, postindustrial gods in place of the glorious beings of the past. Instead of Apollo riding his fiery chariot across the sky or the goddess of night spreading her cloak of stars, we had created little steel-gray gods with the souls of pirates and craft no more beautiful inside than the bilges of battleships.

Or maybe we were receiving a visit from another dimension, or even from another time. Maybe what we were seeing were human time travelers who assumed the disguise of extraterrestrial visitors in order to avoid creating some sort of catastrophic temporal paradox by revealing their presence to their own ancestors.

I wondered about my description of them as insectlike. Their appearance was actually more humanoid. There were no feelers, no wings, no tangle of legs. It was the way they moved—so stiffly—that suggested the insect world to me. That, and their enormous, black eyes. What if intelligence was not the culmination of evolution but something that could emerge from the evolutionary matrix at many different points, just as wings and claws and eyes do. Primitive creatures have primitive organs.

If, say, some species of hive insect had become intelligent on some other planet—or even here—it might be very much older than us, and its mind very much more primitive in structure.

What would a more primitive intelligence be? I wondered if such a thing might not involve less differentiation from creature to creature than we have . . . with less individual awareness and independence.

If they were a hive, they might communicate as a hive, using a complex mixture of sounds, motions, scents, and even methods as yet unknown. Our knowledge of earth's hive species is very limited. An intelligent hive might as a whole be very powerful, but as individuals, quite limited both in strength and in understanding.

They might have, in essence, a single, enormous mind. Such a mind might think very well indeed—but very slowly. That would certainly explain why they were being so careful in dealing with us. A single human being might be far less wise than they, but also a lot faster thinking.

I had been assuming that any visitors would be vastly more intelligent than us. What if that was only part of the truth? In terms of earthly evolution, man emerged only very recently. Maybe that also means that man is not the lesser creature, but the more advanced one. If this was so, then older, less quick-thinking and flexible forms might view us as quite a danger to them. They might even want to imprison us here in our earth, or do worse than imprison us.

And yet, I did not have the feeling that they were hostile so much as stern. They were also at least somewhat frightened of me. I was certain of that.

In some sense, their emergence into human consciousness seemed to me to represent life—or the universe itself—engaged in some deep act of creation.

Sitting there at that moment, I suddenly realized that all my questions about whether or not my experiences were real were meaningless. Of course they were real. I had perceived them. A more accurate question was, what were they? They were not entirely mental, at least not in a conventional sense. Other people had witnessed their side effects. The light seen by Jacques Sandulescu on October 4, the scampering heard by Annie Gottlieb, and the explosion perceived by myself, my wife, Annie, and my son suggested that whatever happened that night was more than a psychological experience. And the dozens of similar stories being told by such a cross section of the population suggested as well that if a psychological explanation was to be sustained, it would have to be a most radical one.

In one way, as I sat there at my desk in the middle of the night, I found the notion of all this witness reassuring. But in another way it actually seemed dangerous. It was corrosive to my understanding of myself and of reality.

My mind turned, inevitably, to the part of the hypnosis that had covered the apparent event in my childhood. I had no memory at all of being taken when I was twelve, let alone when we were aboard a moving train. And yet people report this experience beginning at all sorts of improbable times, the most improbable being while they are driving cars. If it is some sort of hypnagogic trance, why don't they go off the road, or at least run a light or so during the couple of hours the experience usually lasts?

My sister and father and I did take a trip in 1957 to Madison, Wisconsin, to see my aunt and uncle and their children. On the way back I was horribly sick. We had a drawing room on the Texas Eagle from Chicago to San Antonio, and much of the trip took place at night.

At that moment, I felt as if I had opened the door onto my familiar yard at the cabin and found no grass, no flowers, no tall hemlocks. All that was there was limitless, empty blue sky through which, if I crossed the threshold, I would fall forever.

On that note I gave up. It was pushing toward three

o'clock in the morning and I was finally so exhausted that I could not stay awake. I went to bed. I do not remember any dreams. When morning came, it brought me only one thought: I had to see this through. I had to understand, as far as possible, what was happening to me.

No matter what it actually was, my inner self was reacting to this very much as if it were a real experience with real people. Over the next few days I faced the fact that I had a relationship with them, despite the fact that I could not even be sure they existed.

I wished that I could believe that the experience had never hurt anybody. But I couldn't believe that, not in view of the terrific stress I had been under. Before I underwent hypnosis, the great and agonizing issue was: Did it happen or not? In other words, was I a victim or a madman? This question was literally chewing up my soul. It was intolerable.

The day before I met Budd Hopkins I almost jumped out a window. I am lucky. I found Hopkins, and he found Don Klein. What if I had turned for help to someone irrevocably convinced that I was the victim of hallucination or psychosis and too inflexible even to entertain another possibility? I might have jumped. Don Klein's open-mindedness in the face of the fearful and the fantastic has been instrumental in helping me to learn to live with uncertainty.

Because of the intelligence of the approach, I became stronger. And with my strength there came, beginning in April 1986, a great deal of research. I read thousands of pages of material about the whole phenomenon and all its scientific and cultural implications.

If visitors are really here, one could say that they are orchestrating our awareness of them very carefully. It is almost as if they either came here for the first time in the late forties, or decided at that time to begin to emerge into our consciousness. People apparently started to be taken by them almost immediately, but few remembered or reported this until the mid-sixties. Their involvement with us might have been very intimate, though, right from the beginning.

Many of those who have been taken report very early childhood experiences, some dating from considerably before the first disks were reported after World War II.

What may have been orchestrated with great care has not been so much the reality of the experience as public perception of it. First the craft were seen from a distance in the forties and fifties. Then they began to be observed at closer and closer range. By the early sixties there were many reports of entities, and a few abduction cases. Now, in the mid-eighties, I and others—for the most part independent of one another—have begun to discover this presence in our lives.

Even though there has been no physical proof of the existence of the visitors, the overall structure of their emergence into our consciousness has had to my mind the distinct appearance of design.

Despite what they had done to me, I did not hate the visitors. Because I knew their strength but not their motives, they frightened me. But I wished to be objective about them, even to the point of saying that they could very definitely be something from inside the mind rather than from elsewhere in the universe. That they represented a real, living force seemed hard to dispute. But that this force might be essentially human in origin remained a definite possibility.

Certainly, they did not seem human at all. Under hypnosis I described the approach of the small figure wearing the strange cards and helmet. He stood all tiny and bursting with firmness at my bedside and made a semaphore gesture with his closed fists. I leaped out of bed, threw off my pajamas, and presented myself naked to him and his soldiers, thoughts of "the good army" filling my head. There was nothing human about him, except that he had two arms, two legs, and a head.

Given their general reluctance to expose themselves to us in an open and obvious manner, it may be hypothesized that they are trying for a degree of influence or even control over us, but one that at least presents an appearance of compliance on our part.

About two weeks after my last hypnosis session, the three of us returned to the cabin for the first time since Christmas. Anne still knew little, our son, nothing.

I did not realize until I arrived there that I had entrusted these two other people to this experience without letting them know what might be happening. But what could I say? That it could be dangerous? Was that true? Or should I have told them to decide for themselves?

Probably, but I didn't. Later, after Anne had herself undergone hypnosis and knew the whole story, she said that she would never, ever have turned away from the cabin. And our son—if he was involved there was little we could do except agonize. To hear my own mind tell the story, I had been involved most of my life.

As soon as we returned, I realized how precious our cabin had become to me. One would think that the discovery of something like this in one's life would create absolute panic, and special fear concerning the place where it had happened. At first the thought of the cabin filled me with dread, but by the time we went back I felt more reconciled to my situation. It was as if ignoring what had been happening to me required a very substantial effort. Screen memories and amnesia have to be maintained, and there is an emotional cost.

If mine was not an uncommon experience, it might be that we live in a society that bears a secret bruise from it. But as more of us allow ourselves to see what we have fought so hard to ignore, maybe the bruise will begin to heal.

Standing in our living room, I recalled my own distressed state in, really, the past year or so. How many times I had gone through that house or my apartment in the midnight, opening closets and looking under beds. I was always perplexed at myself, because I did not understand my urgency to investigate—especially the corners and crannies. I always looked down low in the closets, seeking something small. Under hypnosis I remembered that I felt as if something very strange was about to put its hand on me and take me away. It was a most curious feeling, not unlike the mood created by the sound of wind in night trees, or the cast of moonlight upon an uneasy stream.

On the surface I was quite normal, but there was this persistent undertone of fear. I felt watched, even though I knew that I was not actually being observed. I bought my gun. I had the burglar alarm installed. I got the motion-sensitive lights.

What was all this supposed to be for in that peaceful, crime-free corner of the world? Even at the time I'd had trouble rationalizing my fears. Now it was clear that they referred not to conventional burglars and such, but to this furtive presence in my life. And this presence was coming to seem to me to be quite real.

The moment I walked into the bedroom I was struck by what I can only describe as a sense of the true. Being in the place where the events had happened made it essentially impossible for me to separate them from other reality. The whole experience of the twenty-sixth, especially the first few moments, rested in the part of my mind that recalled reality, and was as completely distinguished from dream as any other real memory. Those first moments, when the person wearing the shieldlike cards entered the room, were a memory. I was not in any sort of trance when that happened. I was not asleep. I had been conscious and in a state that seemed in every way to be normal.

Now I stood in that bedroom looking at the door, at the floor, at the bed.

Then I saw, behind the door, an indentation in the wall that had been made by the door being slammed open with substantial force.

Did we do that, or did they do it?

Being in the room again, my memories of December 26 grew quite distinct. I could see the visitors coming, could recall my terror, my paradoxical loyalty to their commands. I remembered how it had felt to be touched by them. They were quick and precise in their movements. I remembered the smell of them, the look of their places, and above all I remembered how it was to be with them. There was fear, awe, even a sort of love.

It was cold in the cabin, and I soon went back downstairs and built a fire. We had supper and put our boy to bed with a story, then I sat in a favorite chair with a glass of wine and stared at the flames, thinking.

I wanted to approach this from the most productive viewpoint possible. Rather than trying to decide who the visitors may be and what they may want for themselves here, I thought to concentrate on my own feelings and thoughts about them. Absent even a scintilla of certain knowledge about them, there really was no other meaningful choice.

I looked at the black picture window across the room, seeing only my own faint reflection. How clearly I recalled sweeping off the porch beyond on the night of the twenty-sixth, going into the air with the little people swirling around me. Remaining in my mind's eye is an image that I remembered under hypnosis of my cabin, of my dark bedroom window and my son's window below it with the night light shining faintly behind the curtains. As I left I felt a sorrow as if I had died, and that is why the moment was buried in amnesia.

Then there was an impression as if all the trees below me were suddenly swept up together in a great mass of flashing trunks and limbs. When they straightened themselves out again I was sitting in the small depression with some guards around me. I still had the thought that somebody was talking to me then, but I did not remember what was said. The voice I remembered was one that had appeared to be inside my head. This may seem like some sort of thought transference, but may also be something more understandable.

The brain is an electrical device. As such it has a faint electromagnetic field and even emits very, very weakly in the radio part of the spectrum. Specifically, it contains a portion of an extra-low-frequency (ELF) electromagnetic field wave at frequency between 1 and 30 hertz. The heart and musculature also develop electromagnetic fields in the extra-low-frequency range.

A more acute technology than our own might be able to mediate mental and physical functioning to a great degree by the use of sensitive ELF transmitters and receivers. It is more prosaic, perhaps, than the magic of extrasensory perception, but it has just enough plausibility to suggest that external control of the mind, and even the implantation of perceptions (hearing voices inside the head), is not beyond the realm of possibility.

I should add here that the earth itself generates a good deal of ELF in the 1 to 30 hertz range. Perhaps there are natural conditions that trigger a response in the brain which brings about what is essentially a psychological experience of a rare and powerful kind. Maybe we have a relationship with our own planet that we do not understand at all, and the old gods, the fairy, and the modern visitors are side effects of it. Admittedly, this idea is farfetched. But what of it? So are all the others.

Sitting in front of my fire, though, I found that I could not concentrate on hypotheses. My mind kept returning to memory. Intellectually, I was unsure about what the visitors were. But my emotional self did not share that indecision. My emotional response was to real people, albeit nonhuman ones.

Hypnosis had made the memories very vivid. When I rose out of the woods under hypnosis, I actually felt the sensation of rushing upward, just as if I were in an extremely fast elevator. I saw the trees below me, standing in the night, glowing faintly with a covering of snow. I can remember seeing them between my knees, getting rapidly smaller.

The space I entered smelled like warm Cheddar cheese with a hint of sulfur. This sulfur odor has been reported by others. There was a discarded coverall lying partly on the bench, partly on the floor. I had the strong impression that I was in a living room of some sort. For the most part, people feel that they are in an examination room.

Sitting before me was the most astonishing being I have ever seen in my life, made the more astonishing by the fact that I knew her. I say *her,* but I don't know why. To me this is a woman, perhaps because her movements are so graceful, perhaps because she has created states of sexual arousal in me, or maybe it is simply the memory of her hand touching the side of my chest one time, so lightly and yet with such firmness.

In subsequent months I found that people who have had the experience often felt that they were familiar with one of the visitors, and usually perceived this person to be of the opposite sex. People generally appeared for help with a single memory, maybe two. When they discovered a continuity of experience they were usually as stunned as I was. I also noticed that the fact that people report multiple-visitor experiences over the course of their lives had been published very minimally, even in the UFO literature. I was virtually certain that my own hypnosis was the first time I had encountered what seemed to me to be a fantastically improbable notion.

It was interesting to me that the idea of multiple experiences would be shared by so many people, given that there has been so little cultural reinforcement of it. Published accounts generally avoided this aspect, because a single experience with the visitors stretches credibility so much that an assertion like this seems impossible to maintain.

That night at the cabin, I found myself thinking about the one I knew, turning her presence over and over in my mind. She had those amazing, electrifying eyes . . . the huge, staring eyes of the old gods. . . . They were featureless, in the sense that I could see neither pupil nor iris. She was seated across from me, her legs drawn up, her hands on her knees. Her hands were wide when placed flat, narrow and long when dangling at her sides. There was a structure, perhaps of bones, faintly visible under the skin. And yet other parts of her body seemed almost like a sort of exoskeleton, like an insect would have.

She was undeniably appealing to me. In some sense I thought I might love this being—almost as much as I might my own anima. I bore toward her the same feelings of terror and fascination that I might toward someone I saw staring back at me from the depths of my unconscious.

There was in her gaze an element that is so absolutely implacable that I had other feelings about her, too. In her presence I had no personal freedom at all. I could not speak, could not move as I wished.

I wondered, before the fire, if that wasn't a sort of relief. I had reported that I had been terrified in her presence. Certainly I remember fear. But was that an accurate portrayal of my actual state? If I could give up my autonomy to another, I might experience not only fear but also a deep sense of rest. It would be a little like dying to really give oneself up in that way, and being with her was also a little like dying.

When they held me in their arms, I had been as helpless as a baby, crying like a baby, as frightened as a baby.

I realized that the extraordinarily powerful states I was examining could lead me in two undesirable directions. First, the sheer helplessness that they evoked created awe, which could lead to a desire to comply . . . and then to love. Second, the fear caused such confusion that one could not be sure how to feel.

Her gaze seemed capable of entering me deeply, and it was when I had looked directly into her eyes that I felt my first taste of profound unease. It was as if every vulnerable detail of my self were known to this being. Nobody in the world could know another human soul so well, nor could one man look into the eyes of another so deeply, and to such exact effect. I could actually feel the presence of that other person within me—which was as disturbing as it was curiously sensual. Their eyes are often described as "limitless," "haunting," and "baring the soul." Can anything other than a part of oneself know one so well? It's possible, certainly. To an intelligence of sufficiently greater power, it may be that we would seem as obvious as animals seem to us—and we might feel as exposed as do some dogs when their masters stare into their eyes.

The realization that something was actually occurring within me because this person was looking at me—that she could apparently look into me—filled me with the deepest longing I can ever remember feeling . . . and with the deep-est suspicion.

I wondered that night in the cabin if it was the sheer impact of the experience that had fixed the image of this being so vividly in my mind, or had communion somehow come alive within me? And was she still here in some sense . . . watching even as I sat before my fire?

As I remembered her I found myself filling with a form-less question. Groping for what it was that perplexed me, I recalled an exchange that now came to seem very impor-tant. I'd had a very distinct impression of her, that she was old. Not just aged, like an elderly person, but *really* old. Why had I felt this? I could not be sure.

I still remember her voice, soft, coming from I know not where, answering me: "Yes, I'm old." When she spoke in my head, there was a lilting quality to it. But when she used her voice, it was startlingly deep to be coming from so slight a creature. It was more than a bass: It sounded like it was booming out from the depths of a cave.

I remembered my protest to her when she reassured me about the operation not hurting me. The sense of help-lessness was an awful thing to contemplate. "You have no right," I had said.

"We do have a right." Five enormous words. Stunning words. *We do have a right.* Who gave it to them? By what progress of ethics had they arrived at that conclusion? I wondered if it required debate, or seemed so obvious to them that they never questioned it.

The fire before me sputtered. I opened the vent on the stove and it obediently flared up again.

Maybe their right came from a different direction than one might think. If they were a part of us, it might be that we granted them the right they assert.

Listening to the crackle of the fire mingle with the tick-ing of the clock, I thought that perhaps I might welcome voices of instruction. After finishing *Nature's End* with James Kunetka, I began to feel strongly that the present world situation was unsustainable. I did not think that the world was actually ending, but I could easily have been persuaded that the biosphere would soon change so catastrophically that an immense amount of human life would be lost.

I wondered if a mind, contemplating terrors such as this, might provide itself with gods, if only to ease the burden of being alone with the fear.

If they were real visitors, though, I wanted to know the ethics behind their assertion of their "right." Of course, we ourselves barely question our rights over the other species on earth. How odd it was to find oneself suddenly under the very power that one so easily assumes over the animals.

I thought of some lowing cows, their bells tinkling on a long-ago Texas evening, or of my cat asleep on my lap back in the city, trusting its little self utterly to an affection that to me was casual, but to Sadie was the center of the universe.

I remembered when my father took me to a slaughterhouse in Fort Worth, and I heard the rumble of panic and saw the bucking backs of the steers and the creamy whites of their eyes. I smelled the slick of manure and urine and blood, and heard the steady crunching of the blows and the blare of the saws.

And at a research institute in San Antonio I saw monkey cages with rows of doctored capuchins, shaved, their pink heads sewn or laid delicately open, and the trembling brain probes and the gabble of noise when the vocalization center of one of them was stimulated for the information of graduate students.

What did the monkey with the needle in its brain think of its observers? Were they gods to whom it submitted itself with a noble passivity because it could do nothing else? I saw monkey carcasses in the dumpster, too.

Try as I might, I simply did not have the feeling that the visitors were applying the same cold ethic to their relationship with us as we did to ours with the animals. There was something of that in it though, very definitely. I had been captured like a wild animal on December 26, rendered helpless and dragged out of my den into the night.

Nor did I feel that they were simply studying me. Not at all. They had changed me, done something to me. I could sense it clearly that night but I could not articulate it.

Later, I thought to myself that they were taming me. Maybe this gradual increase in the intimacy of contact that has occurred over the years has to do with that: They are taming us all.

After the dialogue about rights, the female called me their chosen one and I proceeded to get mad. I viewed it as a ploy and reacted with scorn. She wagged her head from side to side, singing "Oh, no. Oh, no." There was insistence in her voice, and humor.

I distinctly remembered seeing a woman wearing a flowered dress being told this. But where? When? The memory was free-floating, without reference. There was just this woman in a white floral-pattern dress standing before a group of them shouting "Praise the Lord" as she was told she had been chosen.

Maybe what they meant was that we have all been chosen—and we are all being tamed.

Nobody has ever domesticated mankind. We are thus a wild species, as wild as the day we first went howling across the savanna. Perhaps the self-taming process of becoming a civilized species did not tame us to visitors, but only to ourselves . . . and then not very well, given our violent history.

That first night back at the cabin, I looked at the couch where they had left me on December 26. I wondered if the old earth did not settle in some obscure, internal way just at the moment I came to consciousness there. Perhaps its low-frequency emissions changed and I fell not from that hidden room in the sky but rather from some lurching walkabout in my own night house. I wondered if there was any relationship between my experience and the mystic walk of the shaman, or the night ride of the witch.

I had read far in the works of mystical search and mythology, and in retrospect it surprised me that I would be so amazed when I finally reached down into the darkest part of the soul and found something there. Now that I was back at the scene of my experience, I felt that I always knew what I would find, and that all of my surprise was itself a sort of illusion.

I reflected that the abduction to a round room had a long, long tradition in our culture: There were many such cases in the fairy lore. The story called "Connla and the Fairy Maiden," as collected in Joseph Jacobs's *Celtic Fairy Tales* (Bodley Head, 1894, 1985), could with some changes be a modern tale of the visitors.

As suggestive as this was of the possibly historical roots of the experience, it was no more definitive of that origin than the whole texture was of the notion of recent visitors.

Maybe the fairy was a real species, for example. Perhaps they now floated around in unidentified flying objects and wielded insight-producing wands because they have enjoyed their own technological revolution.

Every time one decides either that this is psychological or real, one soon finds a theory that forcefully reopens the case in favor of the opposite notion.

The most difficult part of my hypnotic material was the sudden regression to 1957. How could I explain that, even in terms of visitors? To do so, I had to revise my whole understanding of what my life had been. At the beginning of this chapter I described myself as being deeply upset by that unexpected regression. Well, that was true.

But it was no more than a mild state of unease compared to how I felt after I had made a careful inventory of my past.

FOUR

A child said What is the grass?
fetching it to me with full hands;
   How could I answer the child? I do
not know what it is any more than he.
   —WALT WHITMAN, "Grass," from
                  Song of Myself

THE SKY BENEATH MY FEET

A Journey Through My Past

The Journey Back

The more I thought about it, the less able I was to accept the idea that this had been happening to me most of my life. When Budd Hopkins asked me if I remembered anything in the past, I did mention a few odd incidents. The memory of being taken from the train was not among them.

If I accepted that this happened and that it was buried even more completely than the events of October 4, then what else must I accept? Inevitably, that my conscious life was nothing more than a disguise for another reality. It is easy to speculate about such a thing on an idle evening, but when one considered the terrific intensity of the experience I had remembered, thinking that this might have happened again and again had the potential to shatter me.

Still, I could not simply reject the notion. Why should I? Because it seemed improbable? *All* of this seemed improbable. As an experiment I decided to return to my past and see just what I could come up with. As best I was able I reviewed the years for hints of this material. I wondered, though, how I could ever tell if the seeking and the finding were the same act. Maybe nothing happened on that train. Probably nothing did, and there is no way to tell. I would need some sort of corroboration before I could even begin to entertain it as a serious possibility.

It seemed like a trick of the mind. Then I remembered that hypnosis session, and I thought to myself that the real trick of the mind might be happening now. My memories were so spontaneous, and seemed so vividly real. Not the faintest suggestion was made that I regress to age twelve. And yet . . . I now remembered that row of soldiers sleeping on those tables just as well as I remembered the drawing room of the train we were on.

To protect my sanity, I had to believe that this was a comprehensible thing. If it was contact, then it must be proceeding *somewhat* along lines I could understand. They've been here for a while. Fine. Lately, because I moved to an isolated area, they found me. That I could at least entertain. But I could not accept the notion that they were so totally involved in my life.

I found a photograph of myself during the spring of my twelfth year, which showed me in the uniform of St. Anthony's School in San Antonio. Here was a child so clean he seemed to have been polished along with the brass crossed rifles on the collars of his uniform. The picture is inscribed: "For my dear father with love, Whitty."

The neatness was a total deception. It couldn't have lasted more than the precise amount of time it took to snap the picture. At twelve I was usually involved in mischief of one sort or another. I was rarely clean. I was rarely even still.

I looked into the child's eyes. He did not look haunted to me, that boy just flirting with puberty. In May of that year my younger brother had been born, and the house was consequently in upheaval, only some of it pleasant. I spent much time in my room reading. That summer I read *Life on the Mississippi* and it was also the summer of my discovery of Kafka. One afternoon I found my mother reading *The Metamorphosis*. After that I read *The Trial*. I'd go down to the San Antonio Public Library on the bus and sit in the big reading room under the fan and read Kafka until the librarian started getting uneasy, then I'd shift to Robert Benchley for the balance of the afternoon.

My smiling face hid a person full of conflicts, trying to cope with the sudden presence of an infant in an established home and discovering under the sheets at night that the sins the older boys whispered about were real, and were they ever sins!

I was deeply conflicted about my Catholicism, wondering whether the tenets of my faith could be fitted to the picture I was forming of the world. I asked why the pope hadn't saved the Jews from Hitler. I asked why the Church had burned people at the stake, and what on earth did abstaining from meat on Friday have to do with getting to heaven? And if the worst punishment in hell was to get a glimpse of heaven and not get to go, then what about the nuns in Limbo who were there caring for the unbaptized babies the angels didn't want to bother with? They'd had more than a glimpse of heaven. They'd been there for a while. So wasn't sending them to Limbo actually sending them to the depths of a personal hell?

The pope closed Limbo before we worked that one out in catechism class, unfortunately.

Still, my faith was a burning fire in me. I loved Christ and Mary especially, and used to pray with great fervor whenever I was trapped into going to church. Then the priest would invariably say, "Go, the mass is ended," when there were still ten minutes left. But why?

At home I got hold of a book by George Gamow about relativity. Suddenly I understood how the nuns could take Limbo. I understood why the mass did and didn't end at the same time. It was all relative. Einstein, in describing the physical universe, had also described the internal logic of the Church, enabling me to preserve my faith.

But when I brought up Einstein with my mother, she said, "We are Catholic. Catholics are absolutist." She and I would spend hours together sitting on the front-porch steps talking. We discussed everything from general relativity to the price of tennis shoes. I used to try to talk her out of her religiosity, but she was a Catholic intellectual in the heady days of the fifties, when the mass was still full of mystery and there were many fascinating and subtle potentials for sin.

My catechism class was asked to write essays proving the existence of God. Mine, an equation with an intentionally tautological argument, was declared to be a demonic inspiration. When confronted with this by the teacher, my mother said, "To think that children might be inspired by the devil is itself demonic inspiration." Those times were not more innocent than these, but they were less complex.

I do not recall thinking or talking at all about extraterrestrials. However, when I recently asked a friend of those days what was the strangest experience he could remember, I was surprised to find that his answer involved me. At the time I asked him the question, he had not in any way been exposed to this material.

Here is the story he recounted. When we were thirteen I apparently announced to him that "spacemen" had taught me how to build an antigravity machine, which I was constructing in my bedroom. This was in the summer of 1958. I do not remember the genesis of this machine, but I certainly remember building it. There was no magic to the thing; it was only an assembly of electromagnets taken from old motors. The supposed antigravity effect was based on a principle of counterrotation.

When I plugged my assemblage in, there was a great buzzing, the electromagnet in the core of the thing whirled madly, and the lights in the house began to pulsate. The whole thing whined and fluttered. There were showers of sparks. Parental cries of alarm rose from downstairs. As the machine destroyed itself the pulsation of the house lights became a dimming, until the bulbs glowed orange-red. Then they burst to blazing life, a good number of them blowing out in the process.

Finally I managed to pull the plug. Rather than tell my parents what had happened, I rushed downstairs and pretended ignorance. I did not need to pretend fright. The friend reports that I called him in great anxiety and said that I was afraid that the spacemen were mad because I had disturbed their power field.

I have subsequently discovered that there is a whole mythology of flying saucer technology, and a lot of it revolves around the concept of counterrotating magnets. One among the other people I have met who have remembered being taken tells an interesting story. He knows a man, another victim, who was given detailed instructions about how to build a motor of this sort. The man was given the instructions during an abduction experience during the fifties, and claims that he was told that he wouldn't remember a thing until 1985, when he suddenly found his mind full of richly detailed plans.

The exact sizes of the electromagnets and their distances from one another were explained, and there was much about the materials to be used. Not having seen these plans, I cannot evaluate them other than to comment that the idea that counterrotating magnets of any kind would produce any unusual energies at all flies in the face of modern magnetic theory. But he claims that when he built this device, all the metal objects in his barn were instantly pulled toward it and he was knocked out by a flying automobile engine. The next day the barn burned to the ground in an unexplained fire.

It would be easier to believe in the truth of all these effects if superconducting coils were used instead of electromagnets. It is awfully hard to see where a field that powerful would be coming from, given our present understanding of magnetism.

I don't really think that details like the construction of a motor can be part of some shared hallucinatory system. Recall that I did not even remember my antigravity machine myself, but rather was told about it by a friend who remembered. My machine was built in 1958. More than twenty years later this other man seems to have built a more exact version of the same thing, allegedly based on plans obtained in the same period.

The day after I built my device, I do remember being seized with a fierce urge to get away from the house. I went to my grandmother's country home with her, even though the occasion was one of her afternoon card parties.

About four the telephone at the country house rang. I can remember my grandmother saying, "House burned down? Mary Strieber's house burned down?" The blood just drained from me. Fortunately the entire house had not burned, only the roof over the wing containing my bedroom. The fire was never satisfactorily explained, although I have a feeling that it was related more to the effect a little boy's antigravity machine had on the wiring than to the hostility of annoyed visitors.

Fortunately for me, it never dawned on my parents that I might have caused a disaster on this scale.

In July 1957 my father took my sister, who was then thirteen, and myself from San Antonio to Madison, Wisconsin, to see his sister and her family. We flew to Chicago and stayed at the Hilton, where I accidentally dropped a large milk shake out of a tenth-floor window. We spent the night at another hotel, and then traveled on to Madison to see the relatives. A week later we returned to San Antonio on the train.

All my life I have had a memory of that train, seen from above, rushing through the night. Most of the windows are dark, which suggests that it is very late. There are thick pine woods, meaning that it must have been in Arkansas or farther north, for the Texas Eagle did not go through the pine forests of East Texas, but rather across the plains between Texarkana and Dallas and then south over rolling, featureless country.

For some reason I had never thought twice about the strange image of the speeding train. Why would I have seen it from such a position? Can it be that I really was outside of it at some point?

I remembered absolutely nothing about being taken off the train. There was a sort of confused recollection of my father crouched at the back of an upper berth in our drawing room, his eyes bulging, his lips twisted back from his teeth. But I've always assumed that was a nightmare brought on by the fact that I was so sick on the trip. My illness was violent. I vomited until I thought I would die, and for no apparent reason. Nothing came up but bile, but the spasms simply would not stop.

Now I have added to this recollection a vivid memory of the being pushing a bladder down my throat. This is not the only recollection I have of being made to eat things by the visitors. In 1968 I ended up with four to six weeks of "missing time" after a desperate and inexplicable chase across Europe. This is associated with a perfectly terrible memory of eating what I have always thought was a rotten pomegranate, which was so bitter that it almost split my head apart. A nurse put drops on my tongue to help me keep it down. But what nurse? Where? I was never in a hospital.

Something might have been stuffed in my mouth on the night of December 26. I certainly remember them trying to get it opened. And afterward I brushed my teeth.

This is about the most disturbing thing that I have yet come across in this whole, vast experience. It is not the eating that disturbs me, because I seem to have lived, but rather the structural coherence of the thing. First I am fed and it comes back up. Then I am fed again and this time drops are used to prevent the material from returning. Years later, the feeding is such a minor part of the experience that the memory of it is covered by other things.

In short, my hallucinatory friends seem to have learned something about how to get me to digest whatever it is they are trying to feed me.

I have not thought of those hours of sickness on that train for a long, long time. I remember, though, how my father labored to help me, and after he grew tired the sleeping-car porter came in and held me over the toilet. A doctor appeared in his bathrobe and tried to get me to drink some water. The illness had begun suddenly in the middle of the night and continued until morning. I was sleeping like the dead when the train finally pulled into the old MoPac station in San Antonio. My father carried me to a cab, and we went home.

By the time I got there I was feeling much better and was eager to see my friends. We had, after all, been away for nearly two weeks in the middle of the summer, and during the last summers of childhood I sensed the increasing rarity of the days. I was full of excitement as we drove down Elizabeth Road. No sooner had I gone through the motions of helping with the luggage than I was off, my sickness forgotten.

I remember it was then that I told a story, which has remained in the back of my mind for years, of hearing a wolf howling and seeing one on the roadside. Even as I told that story I remember being a little confused. Since then it has lingered, the image of the wolf in the clearing and the sound of its voice echoing through the night. From that image there has flowed an intense lifelong interest in wolves, which has grown into love for this wonderful species.

The image was central to *The Wolfen* and *Wolf of Shadows,* and appears again in *The Wild,* a novel I have written but have not yet published.

I knew even as I spoke that we hadn't really seen a wolf, or heard one howl. Why then was I saying it? Where had it come from? Was it one of the screen memories which were so common to experiences with the visitors? My memory of the December 26 incident was at first blocked by the recollection of the owl. I saw an owl once before, too, during the events of 1968.

I note in passing that if my wise and determined friend from afar is a woman, it could be said that her personal symbol is an owl. Athene's symbol was the owl. The Latin word for owl is *strix,* which also means witch. It was thought in earliest times to embody the wisdom of Ishtar, the ancient Mesopotamian "Eye-Goddess" with the huge, staring eyes. The owl was also the totem of the Celtic Blodeuwedd, the Triple Goddess of the Moon, and is associated with the notion of the Trinity, which will emerge later in this book as the most common symbolic structure of the visitors, mentioned by many people who have been taken—people who have no idea at all of its ancient importance, which has now declined to the dusty precincts of antiquarians and mythologists.

Perhaps visitors would naturally seek to the center of the soul and enter its reality, being too experienced to be interested in any but the deepest essence of our beings. Then they might well seem to be part of our mythology, part of the basis of being human.

My life is full of peculiar stories like the one about the wolf and the ones about owls. Oddly enough, my sister also has a strange story about an owl. Sometime in the early sixties she was driving between Kerrville and Comfort, Texas, well after midnight. She was terrified to see a huge light sail down and cross the road ahead of her. A few minutes later an owl flew in front of the car. I have to wonder if that is not a screen memory, but my sister has no sense that it is.

Many of my screen memories concern animals, but not all. I remember being terrified as a little boy by an appearance of Mr. Peanut, and yet I know that I never saw Mr. Peanut except on a Planter's can. I said that I was menaced by him at a Battle of Flowers Parade in San Antonio, but I now understand perfectly well that it never happened. For years I have told of being present at the University of Texas when Charles Whitman went on his shooting spree from the tower in 1966. But I wasn't there.

Then where was I? And what is behind all the other screen memories?

Perhaps on some level I do know. Maybe that's why I spent so much time peeking into closets and under beds. If I really face the truth about this behavior, I must admit that it has been going on for a long time, although in 1985 it became much more intense. Now that I have uncovered these memories, though, it has ended completely.

As a matter of fact, I cannot remember a time in my life when I have felt as well and as happy as I do now. Whatever has happened, one thing is certain: A great pressure has been relieved, and that pressure had been with me always. Was it the pressure involved in keeping my memories of them hidden? I just don't know.

This brief review at least suggests that I ought to continue my exploration of the past. Before going on, though, it might be wise to examine those last few minutes of the hypnosis session in some detail. It began when, without warning, I found myself in 1957.

Spontaneous regression can happen in hypnosis. The reason it usually tak·s place is that the subject encounters a memory of something that has also been seen long ago, and drops back to the previous experience.

For me the trigger seems to have been the "You are our chosen one" speech. After that, I mentioned "others." At that instant there was a flashing image of the lady in the flowered dress being given some sort of elaborate speech, and shouting "Praise the Lord" when she heard it. Then Dr. Klein asked what others, and I found myself in yet another place but with the same being still before me, or somebody who looked very much like her.

I was excited, sitting up in bed, looking around at the other beds, all of which contained American soldiers lying down asleep. These beds were really more like tables with solid bases and a slight inward cant from bottom to top. I remembered them as being gray in color. The soldiers were young men in fatigues, and they were sprawled as if totally comatose. It was then that Dr. Klein asked me my age, and I heard myself say, "Twelve."

I changed completely, remarkably. I was my childhood self again. It was quite wondrous. I felt smaller, I felt very different. My mind felt different. Gone was the weight of knowledge. For those few moments I was innocent again.

I knew where I was and I was very excited to be there. At first I was sitting up, happy to be awake because even the soldiers were asleep. I was quite pleased with myself. There was no apparent transition between the time I was sitting up awake and the time I was in a little chair, sitting before a featureless gray surface.

Something terrifically difficult happened while I was sitting in that chair. After hypnosis I recalled seeing a landscape with a great hooked object floating in the air, which on closer inspection proved to be a triangle. Then there followed a glut of symbolic material, so intense that even as I write I can feel how it hurt my whole brain and body to take it all in. I don't remember what this was—triangles, rushing pyramids, animals leaping through the air.

Are such experiences the source of the performance anxiety that has been detected in psychological tests I have taken, or does that have to do with the many recollections I have always had of sitting in the middle of a little round room and being asked by a surrounding audience of furious interlocutors questions so hard they shatter my soul? Trying to cope with these memories as a child, I wove anguished fantasies around the figures, who became my childhood friends in some round, gray basement, drawing out the secret structures of my mind like surgeons with forceps extracting sparking neurons from my brain. I remember that they would say words, and each word they said would go through me like a hurricane, evoking every memory, thought, and feeling associated with it. This would go on for hours and hours until I begged them to stop, and I would be offered the relief of a brief rest at their feet, my soul confessing itself into the stern softness of their love.

Is this just a fantasy, or is it what happens when somebody tries to extract the deepest sense of a language from the mind of a child? If so, who did it? Is this a memory of the visitors at work, as it were?

In my childhood I was known as an extremely persistent questioner, so much that in school I was allowed to ask no more than three per period so that I would not take up all the time. True to form, I started questioning this being. Maybe I knew her even then—certainly I had gotten over any initial shock quickly.

I find the exchange fascinating. I asked who the people around me were, and was told what was obvious even to me. They were all soldiers. Then I wanted to know why they had been brought here. The answer is telling: "Because they were alone." It might suggest a methodology, one that is borne out by some studies of unexplained sightings: The craft seem to favor isolated areas. They do not appear as often over cities, and there are not many stories of their taking people from heavily populated places. Perhaps a limitation of technology is visible here: There are simply too many risks in populated areas.

I asked what was being done to the soldiers. The answer, typically uninformative, was, "We look them over and send them home." I can recall my perplexity at that moment very well: I relived it during hypnosis. It seemed an awful lot of trouble to go to just to examine people, and I asked, rapid-fire: "What's the point of that?" The creature seemed ready to reply, but she was cut off in midsentence as if somebody had flipped a switch. For a moment she sounded like a stuck record. "The point of that is— The point of that is—" Then she stopped, as if surprised that she had been caught off guard, and said, simply, "Well," her voice melodious with amusement.

Soon after, I was watching her moving around. I did not know what she was doing, perhaps something that involved touching the soldiers with a copper-colored thing. I asked her why she looked so awful, and she certainly did look awful. I cannot imagine why I wasn't terrified. It is incredibly upsetting to see something that is clearly not human walking and moving about with intelligence. There is something that is unmistakable about the precision of consciously directed movement that is deeply frightening when seen in such an alien form.

Nothing like this was going through my mind at the age of twelve, but the vision of that eerie being moving about among the tables remains quite clear. When I asked her why she looked so awful, she replied almost absently, without stopping her work, "I can't help that."

I wonder what Ishtar really looked like, and if the whole Greek pantheon of beautiful gods and goddesses was not something akin to the beautiful "godlike" beings imagined by people who have made flying saucers their religion. These believers seem to be people who cannot face the stark reality of the visitor experience, and so cloak the fierce, limitless eyes, the bad smells, the dreadful food, and the general sense of helplessness in a very human mythology.

I wonder if Homer and Pindar did not do the same. And why was Homer blind? It is known that many different storytellers comprise "Homer." Perhaps hysterical blindness was a commonplace among the prehistoric Greek bards out of whose tales the classical pantheon emerged. I don't blame them. Hysterical blindness and congenial belief systems would both be excellent defenses against things similar to what I have seen.

But if they have really been here so long, why did they have so much trouble getting me to keep down my feeding? Perhaps the substance has not been changed over the years to suit me, perhaps the very act of eating it has changed me. Maybe it is a process of acclimatization.

It was while I was watching the lady with the eyes moving among the soldiers with her copper wand that I noticed my sister. She was below me and to my right, lying sprawled in her nightie. I still remember how much seeing her like that scared me. She and I were very close in those days. I loved her and admired her, and it was dreadful to see her looking as if she were dead. A voice told me that she was all right. This voice was definitely inside my head, I remember that quite distinctly.

Then I saw the sight that has brought me more fear than any other so far: My father was standing near my sister in blue pajamas, his arms dangling at his sides, on his face a look of surprise. Then his eyes moved until they rested on something I could not see very well, because it was invisible beside the doorway. Almost in slow motion his face simply broke up. He threw his head back and something like an electric shock seemed to go through him, making him spread his fingers and shake his arms. His eyes bulged and his mouth flew opened. Then he was screaming, but I could hear it only faintly, a muffled shrieking, full of terror and despair.

The "awful-looking" creature now came to seem absolutely monstrous. And there was no question in my mind about its being real. It had never even crossed my mind that I might be dreaming. This was as real as any other event in my life, despite the fact that it was far more frightening even than the most frightening horror movie and would soon disappear into amnesia. As a matter of fact, it would be another year or so before I would see my first horror movie, *The Creature from the Black Lagoon,* which was shown at my summer camp. I remained at that camp exactly one day. It was later that my interest in horror stories began. Until I was about thirteen, my taste in comic books ran to *Uncle Scrooge McDuck* and *Little Lulu.* The scariest things I was exposed to were *Alfred Hitchcock Presents* and *The Twilight Zone* on television.

Under hypnosis the fear went through me like ice water in my veins. It is fortunate for Don Klein and Budd Hopkins that I was under the suggestion not to scream, because they would have heard terrible screaming, I am afraid. As it was, the sensation seemed to explode through me. For a moment I thought I was fainting. I remembered that as a little boy I just shriveled up inside to see my father in such an extremity of terror. In those days he was very much my hero. I tried to talk to him, to reassure him that it was all right. He gasped, "It's not all right, Whitty, it's not all right!" and tried to make a grab for me and my sister. His arms came up and just hung in the air while he writhed and his face worked. When he started screaming again, he became muffled.

They watched this with their steady eyes, like huge black jewels.

The closest thing I have been able to find to an unadorned image of these beings is not from some modern science-fiction movie, it is rather the age-old, glaring face of Ishtar. Paint her eyes entirely black, remove her hair, and there is my image as it hangs before me now in my mind's eye, the ancient and terrible one, the bringer of wisdom, the ruthless questioner.

Do my memories come from my own life, or from other lives lived long ago, in the shadowy temples where the gray goddess reigned?

Perhaps the visitors are the gods. Maybe they created us. Robert Crick, the renowned discoverer of the double helix, has postulated that the genetic structure of life is so intricate that it seems designed.

According to studies led by Dr. Allan C. Wilson of the

University of California, there is genetic evidence that the entire human species arose from a single female in North Africa between 140,000 and 280,000 years ago. In other words, it is conceivable that we all started from the womb of a single woman.

If we could slip back in time and find her dashing across the ancient Mesopotamian savanna, would we also find Ishtar gliding above in an enormous triangular spacecraft like the one that was seen over Westchester County in 1983, as elusive to her struggling little creation as flying disks are to the air force?

Or would we find an even greater mystery, that the whole pantheon of our reality was somehow contained in the wobbling mind of that creature, who fell down to thank her raw new gods after a panther leaped at her throat, and by a miracle missed devouring us all.

The Journey Continued

THE SIXTIES AND THE SEVENTIES

This expedition seemed to have reached the edge of a cliff, one that dropped far into the shadows. I wondered, if I really looked, what might I find there?

What had my life really been, and how many other lives have been lived like mine, skidding the surface of this dark mirror? I wondered, in early 1986, if a couple of recent strange events might have some relevance to this inquiry into the past.

During the third week of March I had a very peculiar thing happen to me. Sometime in the night of March 21 at the cabin I awoke and found myself unable to move or even to open my eyes. I had the distinct impression that there was something in my left nostril, and that it was being slowly moved far up my nose. When I tried to struggle, I heard a pop like an apple crunching between my eyes. The next thing I remembered, it was morning.

I noticed during the day that my nose hurt. There was a little bleeding, but as my wife and son had reported similar injuries (without the memory of something being in their noses) the week before, I assumed that it was the result of a head cold and dry winter air. But I never came down with a cold.

After March 15, they'd both had episodes of nasal bleeding, and little knots in their nostrils. Specifically, Anne had a knot in the top of her right nostril, my son in his left. I now developed one in my left nostril. Theirs had gone away without incident, but mine bothered me and I made an appointment with my doctor to get my injury examined. He looked at it on March 26 and diagnosed it as a scratch to the nasal mucosa that had led to the formation of the knot. He correctly predicted that it would subside on its own over a period of days.

I thought no more about this until July 26, 1986, when I received a letter from Donald Klein in which he mentioned that many of my symptoms were consistent with an abnormality in the temporal lobe, and that the method of testing this involved a nasal probe.

The temporal lobe is arguably the most important part of the brain, the seat of "humanity" itself. It is in the temporal lobe that sense is made of perceptions. Structures there are important to emotion, motivation, and memory. People with temporal-lobe epilepsy report déjà vu, unexplained panic states, strong smells, and even a preoccupation with philosophical and cosmic concerns. They also sometimes report vivid hallucinatory journeys.

Initially I seized on this as a way to explain my whole experience. I read some work by Dr. Michael Persinger of Laurentian University that postulated "temporal-lobe transients" to explain psychic and religious experiences. But the more I studied temporal-lobe disorder, the less it seemed an answer. It did not explain the overwhelming sense of the real connected with my experiences. It did not explain the physical consequences. It did not explain the witnesses. Nor did it explain the experiences of others. Temporal-lobe epilepsy was nothing more than another speculation, essentially no different and no more supportable than the visitor hypothesis or any of the other hypotheses that I had put forward. I had to fall back on the truth: I did not know what was happening, but it certainly appeared to be happening in the real world.

As it happened, the week after I received Dr. Klein's letter I met a woman who has had the visitor experience; she began her story by saying that the visitors inserted a probe into her nose, which made a sound "like an apple crunching," between her eyes. She had even drawn pictures of the probe and of the entity that had inserted it. The probe was a businesslike affair, a needle with a small, knifelike handle. The entity was familiar to me because I had seen such beings also.

I then asked Budd Hopkins for information, from his cases, of reports of intrusions into the head. Of his hundred cases, four including me reported intrusions in or behind the ear, three under the eye, and eleven, again including me, up the nose. By far the largest number of intrusions were into the nostril, right into the olfactory nerve with its connection to the deepest core of the brain—and behind that nerve, the temporal lobe.

For what it may be worth, my son and I, who had injuries to the left nostril, are also left-handed. Anne is right-handed.

If the temporal lobe is being entered, then it may not be possible to decide between the temporal-lobe-epilepsy and visitor hypotheses. It could easily be that the visitors are affecting the temporal lobe in such a way as to induce abnormalities that would later be diagnosed as epileptic conditions. As the reading of temporal-lobe electroencephalograms is such a subjective business, I decided that I would arrange two separate temporal lobe tests, one by a neurologist recommended by Dr. Klein, and another through a different psychiatrist. This second neurologist carried out the same preliminary examination done by Dr. Klein's man and came to the same conclusion: There was no evidence of abnormality. He was then given a version of my December 26 story, but we never discussed hypotheses at all. That way, he felt no need to defend his findings against one hypothesis or another. I went to a lab, took chloral hydrate, and endured the insertion of electrodes deep into my nasal cavity. A few days later the results came back: absolutely normal temporal lobe function, confirmed by both neurologists.

So whatever the visitors did, they did not damage me in a way detectable to our science. And I am not a temporal-lobe epileptic. The temporal-lobe-disorder hypothesis was now triply weak: The physical consequences of what happened and the witnesses mitigated against it to begin with, and now the temporal-lobe EEG suggested that I was not an epileptic. Moreover, the brief "transients" postulated by Dr. Persinger could not account for the elaborate experiences I had undergone. Only a full-scale epilepsy could account for them.

So far no hypothesis would explain the motive of the visitors—or the self-confidence they showed by inserting their probe through my germ-filled nasal cavity and into my brain. No doctor would ever do that, which also means that these are not buried childhood memories of operations. There is no operation that proceeds as the visitors do, jabbing their needles up the nose. What's more, the nasal intrusion is not an epileptic prelude. Mine did not occur until weeks after I had remembered and reported my first experience. Far from suggesting a disorder of some sort, the consistency of the stories and the reported side effects—nosebleeds and nasal damage—were a strong suggestion that something real was happening.

Had the temporal-lobe intrusion initiated my experience, I would be tempted to suggest that perhaps all my perceptions were somehow tied to it. But the intrusion did not initiate the whole experience. It may well have profoundly altered my perception of what happened to me—and all my past memories as well. Perhaps that is what it was meant to do.

I thought back over the previous few weeks. Most of the things that had happened since December were well documented, in the sense that I had immediately told others as soon as I was aware that they had happened.

Besides the visitation of March 15, which I will discuss in detail in a later chapter, there was one earlier incident that is worth recounting, because it was this incident more than any other that opened the door to the past. And it did this via my sense of smell. Again, it happened before the apparent temporal-lobe intrusion, not after.

The night of Friday, February 7, we spent in our apartment in the city. I was absolutely frantic. I had an awful feeling. I felt their presence. It was palpable. Most upsetting, I could smell them. I could smell a distinct odor as if of smoldering cardboard, and it was familiar from the past. My wife could also smell this odor; it was one we had both smelled many times. Until now, though, I had not understood its significance. There was also another odor, as if of cheese and cinnamon, that I remembered from December 26.

I remained lying in bed, sweaty and sleepless. But I was shocked to discover that four hours had passed without my noticing, very suddenly. I was reading at midnight, turned a page, and saw by the clock that it was four A.M. and I was no longer wearing my pajamas.

When I got up the next morning I found two little triangles inscribed on my left forearm. I don't know what happened, and there is no way at all to explain the event in a conventional manner. The larger triangle was quite straight, delicately incised in just the outer few skin layers as if by the work of a skilled master surgeon. The other triangle, very tiny, was pointing at the larger one.

On the morning of February 8, I stood looking down at those triangles with the shower pounding on my back. I also remembered the odors I had smelled the night before. Odor is an excellent trigger of memory, and the odor of smoldering seemed to unlock a lot of doors.

I last smelled it in 1972 or 1973. My wife and I had gone down to San Antonio to see my family, and we were sleeping in my sister's old bedroom on the second floor of the house. Across the hall was another bedroom, which had been mine when I was a boy. In the middle of the night I suddenly awoke with the impression that I'd just heard a loud noise. I decided to get a glass of water. As I left our bedroom I noticed a strange smell, like smoldering cardboard.

As I went toward the bathroom to get my water, a small, dark figure with a red light in its hand burst out of my old bedroom and dashed downstairs. I was momentarily astonished, but decided that it must have been a family member. The fact that this individual was much smaller than a human being did not bother me in the least, nor even give me pause. Why not? Maybe for the same reason that none of us remembered the events of the night of October 4. Maybe I was led to reason thus.

There are reports of visitors carrying small lights, and the fairy lore contains dozens of instances of "fairy stones" that glowed.

There was no sequel to the appearance of the small figure, except perhaps a family member's comment the next morning that he had had a terrible nightmare. Nothing further was said then, and he does not now remember the incident at all, much less the contents of the nightmare.

I am amazed to think how much of a fugitive I have been. Another individual I have met who has had visitor experiences, a young woman whose story of a disappearing pregnancy is medically documented as not being of hysterical origin, also describes a lifetime of running. "All of my life I wanted to move to New York because of the lights and the people."

So did I. And it turns out that she lives a block from me. We have both been running like mad, and we wound up around the corner from one another. A coincidence? Probably, but the mind seeks for large and subtle designs, images in clouds, hunters marching the stars, always for the hidden sense of the world. The same urgency to understanding that drew early man to imagine the constellations in the random spatter of the night sky might draw me to make false connections. And yet, without a general theory of coincidence, how could I know what was finally true? I searched on, deep into my past.

At the age of nine I had been sleeping out with a friend on a lovely Texas summer night when something woke us up in the wee hours, perhaps an owl killing a rat, the stopping of the crickets, or moonset. In any case, we found ourselves awake and deliciously alone in the dark. We went exploring the quiet slips of the night, through our familiar places, the wide lawns and tangled bushes, all transformed by shadows into a new world. The vacant lot behind our house was then an acre of tall sunflowers, taller than either of us boys. We were wandering through these stalks when we heard someone coming toward us. My friend turned and ran. I stood there, then turned and ran as well. When I reached our sleeping bags I was astonished to find him already so completely asleep that I could not wake him up. How could he have gone from running in terror to being dead to the world like that? And why was he still outside at all? Why hadn't he gone running into the house? Again, our behavior was totally at variance with our experience.

He and I also saw a huge object cross the sky one summer night, an event that I have always remembered as particularly strange. I called him after a lapse of twenty-five years. We talked for some time, then I asked about those two nights. I told him nothing specific about my other experiences, nor did I discuss visitors. Of the first memory he said, "We were probably just scared by a dog." He had this to say about the second: "Oh, yes, I remember that thing. It was huge. It looked just like a—well, it was strange-looking. And there was a black car." I remembered that, too. Immediately after the object passed overhead an old black car showing no lights went racing down Elizabeth Road in the same direction that the object had gone.

Were these descriptions of events as they had happened, or screen memories? Perhaps, if great care is taken, a method can be devised of finding an answer to such questions, a method more reliable than hypnosis.

I also recalled flying with some people over the roofs of the neighborhood in a thing like a rubber raft, and waking up on more than one morning with bits of grass and twigs in my bed, as if I had been abroad in the night.

There wasn't anything else even that specific, except for a memory of a terrifying round object hanging in some forgotten babyhood sky, and seeing a crowd of big, gray monkeys coming up across the hillside. Apparently this took place at my grandmother's country home when I was about two, which would have been in the summer of 1947.

From the night at age nine to an event in Austin in September 1967, there were few specific recollections except those that emerged under hypnosis, and none was clear. By 1967 I was attending the University of Texas. In the last week of August I had just rented a new studio apartment and moved back to Austin from San Antonio for the semester when I had an experience I now understand to have been what is known as a "missing time" experience, lasting at least twenty-four hours.

I had moved into the apartment the day before and was sitting on the couch about noon eating a TV dinner when I was confused to discover that the dinner seemed to have hopped from my lap onto the coffee table and gone cold. Now I wonder if there might not have been a period of missing time at that point. I remember getting up to reheat the food and noticing that it was already two P.M. I decided that I had fallen asleep while eating. I put the TV dinner in the oven and turned on the timer to heat it for fifteen minutes. Then I turned back to the oven to check the temperature setting. I was suddenly woozy, my mouth dry, and the sun was going down outside! The dinner was cold again, and I had—and have—no memory of how the intervening hours had passed. I got scared, deciding that I had been the victim of blackouts, and tried to make a phone call for help. It was midnight by the oven clock when I put my hand on the phone. There was no discontinuous memory at all, no sense of being unconscious. One moment the timer showed a little after six and the sky outside the kitchen window was glowing, then I moved toward the phone and the timer showed midnight and the sky was black. It was exactly as if six hours had somehow passed in less than a second. I then began trying to make my way out of the dark apartment. I was terrified. I shook with fear, and I was so thirsty I could barely stand it. The next thing I knew, I was in front of the sink. The water was running and running into a full glass. My watch said four-fifteen. I rushed out the door of the apartment, and found myself in the cool of a Texas predawn. At this point I remembered something of awesome beauty taking place in the sky, which I later told friends must have been a display of the Perseid meteor shower, which was not active then but had been early in August. I drove to an all-night restaurant called the Nighthawk on Guadalupe Street and had a huge breakfast of toast, eggs, bacon, cereal, coffee, and at least six glasses of orange juice. When I described this singular twenty-four hours to Jim Kunetka, who is good at coining words, he invented a name for my state. He called it a "larconic trance." For years we have laughed about the larconic trance, but I am not laughing anymore. There is no evidence that I suffer from any malfunction of the brain. And I was as sane then as I am now.

Some weeks later there was a frightening sequel. I was lying in bed at my grandmother's house in San Antonio, reading *Time* magazine. It was late at night and I was about to go to sleep. In those days I used to stay with my grandmother when I went to San Antonio because my brother, then a teenager, had effectively taken over my old room at home.

Lying in that bed wide awake I had an experience so strange and frightening that I remember it to this day with total clarity. I was suddenly transported back in time and back to Austin a few weeks earlier. I leaped into my car and backed out of the apartment house parking lot. It was night and the windows of the car were closed. I couldn't see out at all. In fact, I could see nothing but the reflection of the inside of the car. I was so blind that I was forced to stop. Something approached the car. I was frightened to see, peering in the window with its face pressed almost to the glass, what seemed almost to be a demon with a narrow face and dark, slanted eyes. It spoke to me in a high, squeaky voice, and I remember saying that we couldn't leave the car out in the middle of the street.

Then I found myself in an agonizing struggle. I was at once in the car, fighting to keep driving away but unable to overcome an urge to get out and go back into the apartment, while simultaneously fighting, in the real world, an overwhelming urge to get out of bed and rush outside. I lay on the bed, flopping like a fish. Then it ended. Contrary to my impression, I did not move an inch. The magazine was still propped up in my lap. I could see my grandmother in her bed in the room across the hall, reading quietly. This terrible nightmare had obviously caused not a stir.

Long into the night I lay with the light on. Toward dawn I slept. I believe now that this was a nightmare memory of an attempt I made to escape whatever unearthly thing happened to me in my apartment in Austin. I was reliving an experience which at the time it happened was so unspeakably terrifying that I still don't recall the actual event, only the dream.

There then began a pattern of running that has persisted in my life until the present. A few weeks later I suddenly became obsessed with the notion of getting away from the University of Texas, out of the United States, of going wherever I could, as far away as possible. I fantasized about living in a nice little apartment in some enormous city. I wanted bustle and bright lights, not the sparse Texas landscape and the starry nights.

I didn't have much money, so I contrived various means of getting enough to leave. I obtained a loan from the Minnie Stevens Piper Foundation in San Antonio to study film at the London School of Film Technique. I earned some money translating Seneca's *Thyestes* into English and converting the translation into a film script for the U.T. Department of Radio, Television and Film. I worked as a camera operator. By January 1968 I had saved enough money and I left for London. I have never in my life been so glad to see the back of a place as I was to see the back of Texas. For years I have explained my sudden departure by saying that I couldn't stand the place after the Charles Whitman sniper incident. The truth was, I could have remained after that incident. It was my secret terror that drove me away.

My first few months in London were bliss. I felt as if a burden had been lifted. The school was fun. I spent a great deal of time in film-history classes watching old movies. My nights were occupied at the National Film Theatre watching more old movies. I met interesting friends. Then, in July, there was another incident. I cannot recall what happened with any clarity. It was simply too confusing, too jumbled. I was at a friend's flat in the King's Road, Chelsea. For years I have described it as a "raid" from which I escaped by "crossing the roofs." What I actually remember is a period of complete perceptual chaos, followed by the confusing sensation of looking down into the chimney pots of the buildings. Then there was blackness. I woke up the next morning in my own place with no idea of how I got there. Whatever may or may not have happened in the flat was never acknowledged or referred to again by anybody who was there, with one exception, which I will recount in a moment.

The next day I decided to leave London for the Continent. I couldn't stand England for another week, not another hour. One of the people who had been present in the flat warned me against going, saying that I would "never come back." I scoffed. It was to be a two-week vacation. He said that he would get a witch to cast a spell to bring me back. I thought, *What superstitious nonsense*. Recently I looked him up and asked him about this incident. He couldn't think why he had acted as he did, although he remembered a feeling of dread being associated with my journey.

I took the train to Italy, second class. On the train I met a young woman and we began to travel together. At this point my memories become extremely odd. If I do not think about them they seem fine, but when I try to put them together they don't make sense. I recall that we went to Rome, but that we spent a few days in Florence on the way. For eighteen years I told the story that I stayed in Florence for six weeks. But when I went there in the summer of 1984 to promote Mondadori's Italian edition of *Warday*, I realized that I had almost no memories of the place. Even so, I placidly accepted this anomaly. For some reason, I left the young woman in Rome and dashed off on the train with no ticket, traveling almost at random. I ended up in Strasbourg, where I saw the cathedral, then suddenly rushed to the station and grabbed another train, a local, that crept across France, ending in Port Bou on the Spanish border. There I took a Spanish train to Barcelona. I was broke, so I holed up in a back room in a hotel on the Ramblas. I can remember nights of terror, being afraid to put out the light, wanting to keep the window and the door locked, living like a fugitive, never wanting to be alone, haunting the Ramblas, grateful for the unceasing crowds. The rest of the memory is a jumbled mess. I am just not certain what happened, except that I lost weeks of time. I remember something about being on a noisy, smelly airplane with someone who called himself a coach, and something about taking a course at an ancient university. I also recall seeing little adobe huts, and expressing surprise to somebody that their houses were so simple.

I returned to London in an odd way, weeks later than I had planned, with no way to explain those weeks. I cannot say how I got back. What I do know is that I found myself outside a hotel at about six in the morning. I went in and booked a room, then slept until noon. After lunch I went to my lodgings and found that my room had been let and my belongings stored in a trunk in the basement. The management was quite put out. They told me that I had said I would be gone for no more than two weeks and had disappeared for much longer. Since I had not kept up my rent, my room had been given to another student on their long waiting list.

At the time, I simply accepted all this, stayed with a friend for a while, then found a flat on Westmoreland Terrace in Pimlico, where I lived until December 1968. If such incidents were a frequent occurrence in my life, I might suspect some sort of trance or fuge state. There are certainly many odd incidents, but they are too variable in their nature to suggest the symptomatic consistency of disease.

I recall little more until the spring of 1977. From 1970 until then, my wife and I lived in a two-room flat on the top floor of an old building on West Fifty-fifth Street in Manhattan. We were happy there, if cramped. Our marriage grew solid there, and we became confirmed in our life together. One evening in April 1977, something so bizarre happened that I still cannot understand why we didn't make more of it. With both of us sitting together in our living room, somebody suddenly started speaking through the stereo, which had just finished playing a record. We were astonished, naturally, when the voice held a brief conversation with us.

The voice was entirely clear, not like the sort of garbled message sometimes picked up from a passing taxi's radio or a ham operator. Never before had it happened, and it didn't happen again. I do not remember the conversation, except the last words: "I know something else about you." That was the end. I was left dangling. We did not completely ignore the incident. I called the Federal Communications Commission. A man explained to me what I already knew, that ham radios and taxis and police radios sometimes interrupt stereos. But a conversation, he asserted, was impossible. Our stereo had neither a microphone nor a cassette deck. It was a KLH, a good and relatively inexpensive model readily available in the mid-seventies. At the time, I'd had it for about four or five years.

A few weeks later I became possessed of an overwhelming desire to move. Anne agreed. There were good reasons: We needed more space, and I'd gotten a nice raise (I was then working in the advertising industry). We could afford a move. By the end of May we were living on West Seventy-sixth Street on the top two floors of a brownstone. All went well there until the next year. In June 1978 something terrible happened in the middle of the night. I have variously thought of it as a phone call followed by a menacing visit, and as a series of menacing phone calls. I do know that I called the police, and they came up and checked out the roof, finding nothing. I remember only looking out our bedroom window onto the roof garden and seeing somebody standing there. Just a prowler, perhaps, but it has always seemed to me that there was more to it than that.

Again without relating the incident to a subsequent sudden desire to move, I almost immediately decided to move to Connecticut. We rented a house in Cos Cob, the term to begin in July 1978. We then left New York for Texas, spending most of the intervening weeks there. We slept no more than a few additional nights in that apartment. Again, we felt we had good reason to move. We had forgotten the horrifying incident, whatever it was, and attributed what in retrospect seems the obvious outcome of panic to a rational desire to leave the city. Because Anne was pregnant, we wanted to get out of our walk-up. It never occurred to us that we were making a radical move to another city almost on the spur of the moment. We were running, but we didn't know it.

We didn't remain in Cos Cob for a full year. In early 1979 I was awakened by the bizarre impression that there were people pouring in through the windows of our rather isolated house. I was terrified. We had a new baby. I remember trying to get to him and that is all I remember. A few nights later we were awakened by the neighborhood filling with terrible screams. Even though we called the police, they never came, and nothing was ever said by neighbors about the shrieking. Within weeks we had left Cos

Cob because we were "tired of the country" and wanted to get back to city living.

An interesting further occurrence of screaming took place in August 1986 in Provincetown, Massachusetts. We were staying with friends. In the middle of the night we were awakened by truly bloodcurdling shrieks coming, it seemed, from above the house. Neither our friends nor anybody we spoke to the next day had any memory of anything unusual happening that night—except for one person. When I asked him if he'd slept well, he said that he'd been awakened by screaming. His house was about a mile from ours. He, also, has had a visitor experience in his past.

In January 1980 we took an apartment on the top floor of a high rise on East Seventy-fifth Street. All went well until September of that year. This episode began when I saw a strange light streak down the night sky. It moved faster than an airplane and left me with the feeling that it had something to do with me. I was deeply and inexplicably moved, and left with an obscure foreboding. In the middle of the night we were awakened by our son's crying. He was desperate, almost wild with terror. I rushed into the living room, heading for his bedroom. I recall the impression of a small, dark figure dashing toward the sliding doors that led to our thirty-third-floor balcony. Then there was a terrific explosion and beads of glass burst out of the pantry. I kept running for my terrified baby, reaching his crib after what seemed an eternity. I cradled him in my arms while Anne rushed through the house turning on lights. Then she took our son and I went to see what on earth had happened. A siphon of seltzer had exploded, so violently that the glass was reduced to beads, to dust. There wasn't a trace of the water that had been inside. Anne cleaned up the mess while I calmed our son. Then we went back to bed.

In November we closed on a co-op and by January 1981 had moved again, this time to our present apartment in the Village. A dozen times I have told a story of being menaced by an old college acquaintance, whose terrifying appearances and phone calls had driven us from our Seventy-sixth

Street walk-up to Cos Cob, then from there to the East Seventy-fifth Street high-rise, and finally to the Village. A part of this myth is the kindly detective who hypnotized me and enabled me to identify this individual by listening to his voice on a tape. Then we put a stop to his game by simply phoning him back after one of his vicious calls. But it didn't happen; none of it happened. It's just a screen memory, like the story of the six weeks in Florence that never happened. (After I realized that I had not actually been there that long, I began to believe *another* story, that I had gone to Russia and then to France, and been caught in the French strikes of 1968—without reference to the fact that they ended two months before I crossed France.) But why do I need these absurd stories? They are not lies; when I tell them, I myself believe them. I don't lie. Perhaps I tell them to myself when I tell them to others, so that I can hide from myself whatever has made me a refugee in my own life.

A year passed in the Village, quite pleasantly and uneventfully. Then came what we called the incident of the "white thing." It took place in the apartment and began with the most down-to-earth member of our family, Anne. One night she woke up screaming and reported that something had poked her in the stomach. She had seen it, too: It was translucent white and about three feet tall. She was greatly agitated. Naturally, we took this to be a nightmare. Nothing more was said about it. Certainly, nothing was said to our son.

The next night at about ten I was sitting up and reading. Anne had just turned over to go to sleep. Suddenly I was struck on the arm. As I turned I saw a small, pale shape withdrawing into the hall. I jumped up and followed it, only to find the hall empty. It hadn't been our son: He was peacefully asleep in bed. Again, Anne and I hardly discussed the incident. When she asked me why I'd gotten up, I muttered something about her nightmare being contagious. The next morning I noticed a distinct bruise on my arm, but assumed that I must have banged into a table or something.

A few nights later our son suddenly began screaming the

house down. I leaped up out of bed and went to him. He was terribly frightened. He said that "a little white thing" had come up to his bed "and poked me and poked me."

Neither Anne nor our son showed any physical evidence of injury.

The next Sunday Anne and I were at a wedding reception. I called home and our baby-sitter's mother answered the phone. She said that there had been some trouble, but everything was all right. Needless to say, we went home, leaving the reception almost before it had begun and incurring the permanent anger of the bride and groom. Something had happened to the sitter. She said that she'd been cooking her dinner when a child in a white sheet had startled her by peering into the kitchen from the fire escape. Only my wife heard this story. I did not. We have tried to find this sitter, but it's been years and we know that she did not remain in the area past that semester. We cannot remember her name. There is thus no way to tell whether my wife remembers the story correctly. We finally realized that there was something weird about the white thing. I have to admit that my thoughts went to Casper the Friendly Ghost. Strangely enough, there are other instances of a similar white figure appearing in the context of the visitors, and even acting very much as this one had acted.

However amused I might outwardly have been about the incidents, within a few weeks I was on the run again. The co-op went on the market, although we once again didn't relate our desire to move to any disturbances. We had decided to move back to the Upper West Side. But this time things weren't so easy. We couldn't get enough for our place to enable us to have as much space in a more expensive neighborhood. We finally quit trying.

In late March 1983 something happened. I walked out to get a breath of air for a few minutes and found myself returning three hours later. Anne hadn't been home and our son was in school, but the experience was so inexplicable to me that I invented an elaborate fantasy of having imagined myself back in old New York for the missing hours. I have told many friends this lovely story. I realized as I thought back that it didn't happen. It was a pleasant cover story, obscuring some other events. The truth is I don't know what happened to me during those three hours. I don't even know if I left the apartment. I was just gone.

We finally gave up trying to sell the co-op and instead bought the cabin, which brings me back to the present.

Emotionally, I have a great deal of trouble with the notions of spaceships and visitors. I simply cannot help it, even though I have a feeling that I might seem to future generations to be obtuse. In view of the evidence, the reason for my reticence is obscure, but it is not so different from the reluctance of most of my friends with scientific or academic backgrounds to entertain the visitor hypothesis comfortably.

This is because the idea of intellectually and technologically advanced visitors who hide their knowledge from us is threatening and infuriating. It suggests that there is something ignoble about mankind, or even that we are prisoners on our planet. Those are ugly notions, and I for one would prefer an empty universe to one that reacts to us with contempt or Olympian indifference.

We human beings have a very natural stake in the value and validity of our species and our minds. And this is doubly true of those whose sense of personal worth stems from intellectual work. If the human mind is second-rate, then so are those who live by it.

On a deeper level, though, I find that I am beginning to become a little more at ease with the idea that the visitors might actually exist. This is for an unexpected reason. I think of those rushing little figures, those haunting eyes, the smells, the little rooms, the uniforms, the sense of hard work being done. I remember how stiff and insectlike the movements of the visitors seemed, and how very careful they were to keep me under control at all times, and I think that I may know the reason for their peculiar manner of dealing with us. If I am right, then the source of their reticence is not contempt but fear, and well-founded fear, too. They are not afraid of man's savagery or his greed, but of his capacity for independent action.

I have seen them from close range, and if I was seeing real beings, then what was most striking about them was that they appeared to be moving to a sort of choreography . . . as if every action on the part of each independent being were decided elsewhere and then transmitted to the individual.

I return to the thought that they may be a sort of hive. If this were true, then they may be, in effect, a single mind with millions of bodies—a brilliant creature, but lacking the speed of independent, quick-witted mankind. If they think slowly enough, it may be that a human being, fast-thinking and autonomous, could be a remarkable threat. It may be that an old, essentially primitive intelligence has encountered a new, advanced form, and is frightened of the potential that our completeness as individuals gives us.

I can picture myself on some night of the future, watching them approach my bed. It will be so dark that I will barely be able to see. But I will see their short selves dressed in their familiar jumpsuits. I will see those big, bobbing heads and those grave, sharp eyes. I will feel their cool, tiny fingers upon me and hear their breathing as they carry me away, perhaps even catch an occasional high whisper, words said and thought like equal sails upon the same ocean. I will know then the reason for both their interest and their shyness: We awe them and frighten them. And I will understand why.

If I am right about them, it is unlikely that there will ever be the kind of open contact between our two species that seems so logical and useful to us. Even a well-intentioned human being would pose a threat, in that his accidentally taking an action they had not anticipated might cause them literally to lose track of him right in the middle of one of their own craft. Might he not then be able to explore it at will, learn its secrets, and potentially, at least, release all of us into the cosmos?

Can it be that any one of us has the potential to be at once inferior and superior to their entire species?

To contemplate such a notion makes my soul ache with longing to know for certain, and yet also to leap up, as if by some obscure hand it has suddenly been set free.

Hypnosis

[Once we reviewed this apparent past material, Don Klein and I decided to go fishing in it. This and the next would be the last times that Don would hypnotize me. Despite the fact that many complex experiences took place between April and October of 1986, our interest shifted after March from exploring new material to discovering some physical cause for it.

We will return to hypnosis in the future, but—while it is obviously of great interest—it does not advance any real understanding of the origin of what is happening, only of its content. Why, as I am no longer terrified, the more recent events should still be difficult to remember in their entirety without hypnosis is unclear.

We decided to cover the night at my grandmother's house, the fall of 1980 in New York, and anything else that might be of interest. We both recognized, and I wish to make this clear, that there might well have been a degree of degradation taking place, in the sense that I might have been unintentionally fulfilling expectations of seeing the visitors. While at the time of the hypnosis sessions recorded here and in the next chapter I was still avoiding reading books about anything connected with this sort of material, simply dealing with my own memories must have affected me, altering and changing my perceptions in ways that are probably impossible to detect. Despite this, there was confirmation from another witness that one of my memories was indeed of an extremely strange event.

At this session, Don and I examined an incident that took place in the country on an October or November afternoon in 1984. I was driving back to the house from the grocery store when I suddenly saw a fogbank. It was a clear fall day, the air dry. I got curious about the fogbank and drove off the highway onto a dirt road to try and get a better look at it. The next thing I recall, I was in the fog in my car and two people in dark blue uniforms were leaning in the windows. Then I was back on the highway, returning home. I had dismissed this whole thing until just recently, when I thought about it and decided to go back down that dirt road. I went to the exact spot on the highway where I had turned off. I remembered it because it had a lovely view, and I had looked at that view just before making the turn. The dirt road I had seen there didn't exist anymore, and there was no sign of a road ever having been there.

Budd Hopkins had wanted me to cover this incident first, before any further hypnosis, because of its similarity to one of the most common abduction scenarios, the removal of the subject from a moving car. He did not tell me what I have subsequently discovered, that confusion of place is common among abductees. There are stories of roads that don't exist, beaches that aren't there, structures that later prove never to have been built. There are also cases where clouds came down to the ground or strange fogs proved to contain something more than droplets of water. One might be tempted to ascribe such reports to trance states, but that does not explain what happens to the victims' cars while they are in the trance—and there are often hours of missing time involved. In addition, many experiences take place while moving cars are filled with people. Some period of time passes, and the occupants all wake up to find the car moving in a different direction, or at the wrong place on the highway, or some such thing. They do not find themselves where they ought to be: in a ditch.]

Dr. Klein: "You're in the car, in the car—"

"I went right past the turnoff. I went right past the grocery store and I keep going. I don't know . . . I want to take the car for a little run. Listening to WAMC."

"What do you hear?"

"It's *Don Giovanni*." (So I said, but it sounded awfully strange.) "Going . . . down the highway toward the interchange. I keep thinking I see something above the car. I'm a little nervous. I turn off the radio. I roll down the window then roll it up again. I don't know why I missed the turnoff, and I'm going to turn around and go back. But I don't. Isn't a soul around. I calm down, I turn the radio back on again." (I remember flipping the switch to find that it was already turned on. It had stopped working—which was typical of the car I had then.) "I get down past the diner, there's a real nice view off to the left. Looking out the window of the car. A white truck goes past. I—it's like the white truck isn't right. There's a—I don't know what is going on here. Now I want to go home. I feel terribly sick to my stomach. Awful feeling. I don't want to tell you what's happening to me."

"Perfectly all right. Just relax. You don't need to tell us. Only if you can. Relax. Tell us what you can."

"I was driving my car, all of a sudden there was this white pickup coming toward me. Funny white pickup with a black windshield. And the next thing I know, I'm just stupefied. I wasn't thinking about—I just wasn't thinking about them at all, and just the damndest thing. I'm sitting there in a long room. And there's this—being—standing in front of me. A long, gray room. Bang. And I jumped down and wanted to get back in my car. I didn't know where the hell I was, what had happened to my car, anything. It was totally immediate. And then there's this—I feel like I'm being stared at. I have the feeling there's a lot going on but I'm just so totally stupefied. I can't describe how I feel right now. It's like I just got turned inside out. And one thing I do feel, which is my stomach feels terrible. I just can't credit—I can't understand it at all."

"Don't try to understand it. Don't try."

"There's this one right in front of me. I'm sitting on the

floor with my legs spread apart. I'm dressed in my clothes. Wearing my brown sweater and I can see my shoes. 'Cause I'm sitting with my legs spread out in front of me. And somewhere there's someone watching me with great big eyes. Big black eyes. Just watching me. One second I think they're mean, the second second, I don't know what to make of it. I'm not scared now. I'm just amazed and to-tally—totally—it's like—just like I—I turned the corner—and all of a sudden I was in Arabia or something.

"And I'm thinking, there was this white truck—I'm try-ing to figure it out. What happened? I was sort of scared because I feel like I've done something wrong or gotten into the wrong end of the tunnel or something. And over here [gestures left], there's somebody who's moved around a lot. No one says a word. I don't say a word."

"Moved around?"

"I could see out of the corner of my eye someone or something is moving around a lot. Just sort of darting around. And there's a whole row of people—little people—standing quietly over there on a little—they're a little higher than I am. And I'm still just—my mind is whirring and whirring and whirring like I'm short-circuited. I mean, it's a—a—on—like I'm on overdrive or something. And I have this feeling that I could kick my way out or dig my way out or something."

"Are they communicating with you?"

"No."

"You're just there?"

"I'm just sitting there."

"Are they paying attention to you?"

"Yes. There's one of them now sitting down in front of me staring right at me, and she's completely different from the others. The others are all very small people. This one is tall and thin. And she's sitting down. She's all gangly. I don't know what to make of that. I don't know what to make of this. Where the hell—how the hell—you know, it's like I can't see. I just don't know what the hell to make of this. It's just impossible. It's totally impossible. It can't be like this."

"Maybe it's not like that."

"How the hell is it, then?"

"Look at it very hard, see if you see any changes. Look at it very hard."

"She's staring right back at me. She looks like a big bug. Just sitting there, staring at me."

"Are you staring back?"

"I don't know what exactly I'm doing. I'm feeling very sad."

"Sad?"

"Sad. Yeah, I'm looking at her. She's looking at me."

"Do you know why?"

"No idea. I just don't understand it. It's very hard to understand."

"You say she looks like a bug?"

"Yeah. Great big black eyes. She's sort of brown. She has a little, tiny mouth. She's thin."

"She have antennae?"

"No."

"She have hair?"

"No, she's bald."

"She have ears?"

"I don't see any."

"Eyebrows?"

"No."

"Does she have a nose?"

"A little bitty tiny sort of two-holed nose."

"A nose, or just the holes?"

"I guess it's there."

"What's the mouth like?"

"It's straight and sort of—it's straight and—for some reason it's a little hard to look at."

"Try to make it out. Horizontal lips?"

"Yeah. It's very slight. Just an opening. Very slight lips. Sitting there like that." (I drew my hands around my knees to demonstrate the position. Then I paused, remembering a confusion of images.) "Something happened to me just then. She sat there for a long time, then put a hand out, put it under my shirt and under my sweater and under me and put it right up against my chest, on the side of my chest. And it felt sort of soft, and it's—it doesn't feel bad to be touched like that by that thing. And she takes her hand away.

"Where the hell am I? I'm way out in the country. I thought I was, uh—you know, I'm just scared to death. I'm totally coming out of it. I'm not out of it, either. I'm wrong."

"Try to relax."

"I'm just scared to death, Don."

"Just relax. Sit back, relax. The fear is real but it can't hurt you. Just relax. When did the fear start?"

"When I realized I was driving down this road and I didn't have any idea where I was. I was in the woods on a dirt road—where—what—how'd I get here? So sick of—I was driving down the highway, then I see this weird white pickup. Then I'm all jumbled up and confused. Then I'm sitting in my car on this road scared to death."

"Do you have any recollection of two people in uniforms being there?"

"I'm just sitting in my car alone."

"Anybody tell you to go back?"

"Yeah. He says to me, 'Get out of here.' Then this lady on the other side says, 'We don't want you here.' I say, 'Who are you?' She looks at me with a real mean look on her face. She's a—real mean."

"What are they wearing?"

"I mostly looked at the one over on this side." (Passenger side.) "I thought that was a woman. You know, I just can't tell what's going on here. I don't know what the hell happened. Because the next thing I know, I'm on the road again. I'm going back home."

"I want you to relax. Relax. Let your body go limp. Relax. Deeply asleep again. Deeply asleep. Stay calm. I want you to report. Be a reporter. Tell me what happens. I want you to go back to that long room. You are looking into the eyes of this person. You said this was a woman. Why do you think it's a woman?"

"Because it is."

"Did you hear her talking?"

"No. She told me a long time ago."

"This is somebody you know."

"I don't know."

"Remind you of anybody?"

"I don't know. Don't ask me."

"Try to stay with it."

"Yes—reminds—somebody—I— [Gasping. Evident severe distress.]"

"Try to go on."

"I can't, because I can't make any sense out of it. It's like there are huge, swirling . . . she's got something, she points it at me, it makes tremendous, swirling pictures in my mind, of—I don't know what it's of. It's not of anything. It's like, uh—it's sickening. It's very—they're pictures of abstractions. Things fitted together. [Pause.] I feel much calmer, much better."

"Why do you feel that?"

"I don't know. Just do."

"What did she do?"

"Because of these pictures."

"They made you feel better?"

"Better. They're abstractions, like triangles and circles and things. And they're fitted together in order. The triangle with the circle in it and the square comes around it and it moves all very smoothly, and it makes me feel better." (Note: When people are asked at random to draw the first figure that enters their minds, 30 percent will draw a triangle. Nobody knows why this is the case.)

"Did she want you to feel better?"

"I don't know. Nobody said anything about it to me."

"This person—you saw somebody when you were twelve years old. Was that the same sort of person?"

"Yeah."

"Exactly?"

"I don't know. Looks about the same."

"You said she was very tall and thin."

"She's always sitting down. She's got a lotta legs and arms."

"A lot?"

"I mean four." (Two legs and two arms.) "But she's so thin, and her arms are especially thin. She has sort of hands that look like they might have gloves on. You know, I've seen her before."

"Go back to when you saw her before. Try to go back to another time."

"When?"

"Whenever you saw her before."

"I've seen her lots of times."

"Lots of times?"

"Sure. I've seen her lots of times. I hate to think about that. Christ. I really do. Really do. It's very hard to think about that. It's like I'm being horsewhipped. I'm just not going to think about that. I don't want to think about it."

At this point Dr. Klein no longer wished to continue the hypnosis because of my evident distress and awakened me. My first words on waking up were, "It was like my head was in a vise right at the end. Like I was being beaten, whipped. Horrible. Why would it be that intense?"

"It seems like there's a lot of bad feeling here."

"It was like a stone. A stone blocking me."

The night after that hypnosis session I became aware of an almost palpable presence before me. It was the image of her. After my careful scrutiny it had lingered past hypnosis into my conscious life. As I walked home that night I could see it before me just as it had been in the long room, staring at me with its great, dark eyes.

I had the impression during hypnosis that I had been going somewhere that afternoon, someplace I wasn't supposed to go, almost as if, having been drawn to buy a house in that particular area in the first place, I was overreacting to the impetus and was on my way right into their laps. Can it be that there is some sort of lodgement or projection into physical reality where these manifestations congregate? Or is it simply that visitors have established a base in the general neighborhood . . . or, perhaps, have been there since time immemorial.

After this hypnosis session I suffered the same sort of debilitation I had felt for so long following the December 26 experience—lowered body temperature, weakness, an unpleasant sense of being somehow separated from the world around me. And the next afternoon I felt terribly tired.

Either I was overloaded by a demand for more material than I could comprehend at one time, or there was a limit to the amount of information I would be able to remember.

Or perhaps I could not comply because my unconscious mind had not anticipated the question, and had not yet had time to construct more stories.

I wonder.

Hypnosis

FISHING IN THE PAST, PART TWO: THE DEEPS
SESSION DATE: *March 14, 1986*

SUBJECT: *Whitley Strieber*
PSYCHIATRIST: *Donald Klein, MD*

[This time we tried for a very, very deep trance. I was concerned about complying not so much with Don's wishes, since he remains neutral, but rather with a hidden desire of mine that my memories would somehow confirm their own reality. I was hoping that a more profound trance would direct me totally toward Don and thus all my compliance would be concentrated on him, and I knew I could trust his neutrality.

The trance, which was deepened by the added suggestion that I descend two flights of stairs of twenty steps each, was profound. When I awakened it was as if I had come up out of a well. I was reasonably satisfied that I had been too directed toward the hypnotist to interfere with myself, but it is not possible to be certain of this. I noticed afterward that he seemed to have kept his questions to an absolute minimum, perhaps in an effort to further reduce suggestion.

We went first to the incident at my grandmother's house in 1967. The images were startlingly real, almost as vivid as if I were there again, twenty years ago.]

"Granny's in the other room. She's reading too. I see her light is still on. Gigi's sleeping beside her bed. The rest of the house is dark. I turn the page. There's an ad for a car. Just staring at that ad. Goddamn. Something buzzing around in here. I'm fighting it. All of a sudden it puts something on my head like a railroad spike. A silver nail. A big, flat-headed silver nail and it hits me right in the side of the head with it. And I turn into something—else. I have—I'm heavy and big. I get up out of the bed, I feel totally different. I feel like I'm moving, like I'm walking through the house, like I'm a ghost in the house. I'm in the basement.

"No I'm not. I—I'm still in bed. It's so peculiar. Because I never moved at all. The ad's still there. I never went anywhere, I stayed right there. I get up and I get a glass of water. I'm scared to death. I don't know why this happened to me."

"You said something about something happening right at the beginning of the thing?"

"Yeah. There's like this very fierce face up beside the bed. A very fierce strange face like a giant fly. It hits me with like a bright silver nail. This huge change comes over me. It's terrifying. Like I turn into another kind of a person. I'm big, I'm heavy, I'm dark. I'm bony. And I start walking out of the room. And it just—it's terrible. And the next thing I know I'm back in bed again, my stomach's all knotted up, I'm scared to death, and the magazine's still on my lap. And then—"

"Could it have been a dream?"

"I don't know. I don't know what to think of it. Yeah, of course it's a dream. It's gotta be a dream. Yet—I don't

155

want to look up there where that thing was. So scared I feel like I'm just totally—it's just totally unaccountable. I don't know what came over me. It's like it came out of nowhere. And there's this face—real mean, fierce face."

"A person?"

"Nah. It's like a big bug. Only it looks at you and you just want to look away. And it has a silver nail in its hand and it hit me with the flat end of the nail and the next thing I knew I was moving. Only I was—there I was back in the bed reading the magazine, still on the same page. An ad for a Mustang very much like my Mustang, only it's blue."

"You had the experience before with that object?" (He probably referred to the October 4 incident, but that was not one that came into my mind.)

"Yeah, it happened to me before. We were out in the lot. Someone came across the lot. My sister thought it was a fireball. I thought it was a motorcycle. We had—they put up the tent, I came out late. I was just gonna go in the tent. It had been raining. All of a sudden this thing came catty-corner across the lot. And they're all stopped. It had a skeleton on it. Scared the hell out of me. I turned around to run in the house, I didn't go anywhere. That skeleton's in the bushes. It knocked down the tent and they didn't do anything, they just stayed inside it. Patricia and Roxy and Angie and that was all. And it had a silver thing, and it had real long arms. I was yelling for Daddy, and I couldn't—there was just nothing happening. I was yelling for Daddy and it comes—I can feel its hand on my shoulder. I don't want it to have its hand on my shoulder, it's really terrible. [Making running movements.] I can't run away. I can't run away. And it has its hand on my shoulder and I can't run away from it. It just—it's stuck to me kind of. Horrible! It pulls itself around me so I can see it and God, whoo! Jesus! [Gasping.] Just looks at me. And then there—she's trying to get this thing on me.

"Oh, I feel much better. It feels—I know it's there but I'm not scared, I'm not trying to run away anymore. This is really something. Boy. I'm lying down on the grass. On the lot. I'm not on the grass, I'm on the lot and it's all stickery underneath. I just had my T-shirt on and it's got something that's gonna—I just know it's there, I can see it clear as day. It looks just exactly like a bug. A praying mantis is what it looks like. Only it's so big. How can it be so big? And I don't feel so bad but jeez, it's just doing that. It's got this thing and it's like working it down into my hair or something. I can't feel a thing. Now it's up in the air. And it's gone. And I know that the stars was coming out and it'd been all cloudy before."

"How old are you?"

"Twelve. And now they're all yellin' at me because I collapsed the tent. My sister's real mad and she's gonna get Momma to spank me because I collapsed the tent. And they get scared. And we take our stuff and we go on inside. Patricia tells Momma she saw a fireball. Mom and Dad are inside watching TV. They're watching—*Ed Sullivan Show*. And Momma says we'll be able to see Señor Wences if we stay inside. And it's cool anyway, cool in the house. It was hot and muggy outside. I feel—Patricia tells Momma there was a fireball in the lot is why we came in. And I wonder why she doesn't say I collapsed the tent. And there was no fireball in the lot. Momma says not to worry about it, it was just a fireball.

"And that night we slept out on the screened-in porch. Daddy came out and sang us a song. It embarrassed Patricia but I loved it. So did Roxy and Angie."

"This vision you had of this praying mantis thing—is that the same as the others you've seen?"

"They all look like that. Yeah. I thought at first it was like a skeleton on a motorcycle or something. It was flying—no, it wasn't flying. I could just see it, and I see it almost does, really does, look like a praying mantis, only bigger. It's got great big eyes that just scare the hell out of you. Scare you real bad. Big, big eyes. Doesn't *really* look like a praying mantis. They've got white eyes. This has got dark, black eyes. I thought it was a motorcycle because at first it looked like a guy on a motorcycle wearing dark glasses. Then I saw, God, it wasn't that!

"It came right up past the tent and just up to the honeysuckle hedge. And I just wanted to get across that hedge and into the yard so bad. Oh, God."

"So you've had this experience of something being done to your head before?"

"Yeah."

"The same sort of thing?"

"No, it calmed me down. It felt . . . good. It tapped me with something in the head and it calmed me down. Felt good. And I knew that it was still there but I just laid right down in the—was I *in* the backyard? No, I was in the lot. But it didn't feel like a bad place to lie down. There's lots of stickerburs in the lot. It was a little stickery but it wasn't too bad. I lay down and looked up at it and it worked something into my hair. I didn't feel anything. It had something long that it worked into my hair. It was like, working in my hair. Doing something. Then it went away. I wasn't scared while it was working that thing up in my hair. I don't know what it was. I didn't ever see it. It was working a thing up in my hair. I didn't feel a thing."

"Try and relax, now, relax. Muscles are loose. Deeply asleep, relaxed. We talked before about another time, where you thought you saw a meteorite."

"Yeah."

"How old are you now?"

"I'm thirty-six."

"Tell me what happened."

"Well, I've started working with the young people at the Foundation." (This was the Gurdjieff Foundation, an organization devoted to the development of consciousness based on the ideas of the philosophers G. I. Gurdjieff and P. D. Ouspensky. It is not related to groups that make public claim to be the "Gurdjieff Foundation." The real Foundation is not a public organization.) "I've been in a very intensive period. A lot of meditation and effort. I'm working very hard on my new book." *(The Hunger.)* "Jim Landis is on sabbatical, so

I'm just working pretty much alone. I feel pretty good about it. We worked together a lot on it. And now he's gone on sabbatical. I think they like it a lot down at Morrow.

"I saw this thing—I had been feeling very good about my Foundation work. I thought I was really connecting with the young people. Maybe I was reaching a new state of consciousness, too. And, looking out over the city one night, the sun had just set and it was beautiful, orange glow; you could see the streetlights everywhere. I loved that den I had then. Loved the den. It's so beautiful a view. When it's clear I can see all the way up to Bear Mountain. When it's clear.

"And then I saw this meteor. Like you couldn't quite really see it. It came down out of the sky, very tiny, like a spark. And I thought, *Oh, that's not a plane. It's not anything.* And I felt awed. Like it was someone coming to see me. I told Anne about it. She said, 'Who would it be?' I said I didn't have any idea. And I went to bed.

"I love her so much. You know, I can't believe I'm seeing this. One two three four five six. I look up from my book and there are six figures standing at the end of the bed looking right at both of us. She's turned over and she's asleep. I say, 'Anne, Anne, look at this.'

"They're menacing-looking. Strange. I don't understand where they even could have come from. They made no sound. They came out of the living room. The door is so dark. Nelson's sleeping under the bed." (Nelson was the family dog.) "'Nelson! Nelson!' Nelson's just sleeping under the bed. I feel like I've just gotten some kind of weight on me. I want to get up, I'm thinking about my son. I want badly to get up. I'm thinking about my son. I do not know what's going on here. Is it because we're so high? Why did I ever get this apartment? Why didn't I get an apartment downstairs? I feel like crying, because I can't get up and the door is so dark. They're just standing there. They don't say anything, they don't even look like they're alive. Five, six, tar babies. Six tar babies standing there. Am I seeing things? 'Anne, Anne!' No one sees this. It must be that I'm seeing things. I close my eyes, I open my eyes. And it's changed. Now they're around both sides of the bed, about halfway up, like when you stop looking at them they start moving.

"I've lost my mind. I have lost my mind. This cannot be real. 'Anne? Anne!' I shake Anne. I don't stop looking at them but I shake Anne. They're wearing uniforms. This just is incredible. It can't be real. It cannot be true, it just can't be. 'Anne?' Why in the world won't she wake up, she's never been this damned asleep. 'Anne, will you wake up! Anne, *Anne!* Oh, Christ.' It's like I'm in another world. I can't make her wake up, and every time I glance at her they get a little closer to the front of the bed. This is really a bad nightmare, boy. This is a real foul nightmare, man! Oh, God, I wish I could wake her up!"

"Voice 1 [Basso profundo]: 'You're not trying to wake her up.'

"'I'm not trying to wake her up.' I feel calmer, better.

"Voice 2 [Light]: 'They're all right.'

"'They're all right.'

"Voice 2: 'They have to do that.'

"'They have to do that. Who's that standing out there?' I get the feeling this place is full of people. Someone keeps telling me it's all right, it's all right, all right. 'I know it's all right, but I still want to get up and uh—ah—'

"They moved again. When I tried to get out of bed they moved again. And they're standing one two three four five." (Total eight.) "Three up right beside my bed, four five at the foot of the bed. One two three on the other side of the bed. And goddamn Nelson is snoring away under the bed. Why doesn't the dog wake up, at least? What's the use of buying a dog? They aren't people, therefore it's a dream.

"Our boy says, 'Oh!' He says, 'Oh!' And he screams out loud, real loud! Screams again! Ah! They pull—all—right down to the foot of the bed and I get up and I'm running like hell and he's screaming like hell, and that praying mantis is standing right in the middle of the living room. Right near the windows. And I run on in to my boy and he's put his arms out and he's got his face all screwed up and he's screaming. Screaming. I never saw that little boy scream like that before. Something happened—something happened to Anne. 'Anne? What the hell was that?' I pick him up. She comes in. He's like he's hard and cold. He's very cold. He's got his diaper pulled down around his knees. It falls off. We're holding him, both of us are holding him. He finally calms down. 'What happened, Anne?'

"'I don't know.'

"'My God, something blew up in the kitchen.'

"'You keep him, hold on now. I'll go. It's a bottle.'

"He's falling asleep in my arms. 'What kind of a bottle, honey?'

"'A bottle of seltzer blew up.'

"'You're kidding.' I've still got my son, I walk out there.

"'Don't get in the glass. There's glass all over the floor.'

"I take him and I rock him. She's out there working on the glass and I'm rocking him. We've got all the lights on. I'm rocking him."

(Don then suggested that I was back in my bed, and that I would be able to see the ones around it very clearly.)

"They look like they're staring at me with their mouths opened. Only their faces do not move. I don't know exactly what they are."

"They're like the other one?"

"No, no, the other one is thin and bigger. They're stocky and little."

"What color are they?"

"Color? They're wearing blue uniforms. Dark blue uniforms. They're sort of gray. They look like they haven't been out in the sun in ten years. Sort of mushroomy-gray. Smell funny, too. Like a burned match head. Just totally expressionless faces. Two big round eyes and a round mouth—and I don't think they even have noses. I really didn't look at them too hard. I don't know if they had noses. I was scared pretty bad there. This has to be a dream, because the dog is sleeping like the dead. Why do you feed a dog?"

Don then brought me out of the trance.

I was shocked by the unqualified reality of what I had seen. I just could not believe, in that moment, that the forms persisting in my mind were anything but real. And yet they had to be something else, surely they did.

There was a certain way of checking the reality of at least one of these memories, because there was one other person involved whom I knew well: my sister. I had been thinking about calling her. I did so the next afternoon, and asked her our now-familiar question: "What is the strangest thing you remember ever happening?"

"The time we were sleeping out in the back lot and the fireball came across the lot."

I sat there holding the phone and feeling as if I were falling down a deep well, and at the bottom of the well was somebody with huge, shining eyes.

"Can you describe it?"

"It was a big, green fireball. It came catty-corner across the lot. We all got scared and went inside. We slept on the porch instead."

"Did we tell Mother and Dad?"

"Mother said it was nothing to worry about, it was just a fireball."

I still do not remember seeing the fireball. All my life I have had a free-floating memory of a skeleton riding a motorcycle, a frightful effigy. Now I know the source of that image.

Were my sister's words confirmation? Yes—of the fact that something disturbing happened in that vacant lot so long ago. But the issue of what it was remains open.

What is most interesting here is really a pattern. It involves two types of interaction with the visitors. One type seems to involve the approach of a single individual or small group, as happened on the night of the fireball, at my grandmother's, in the apartment on East Seventy-fifth Street, and in the country on October fourth. The other type of incident is the long visitation, as the ones that occurred when we were on the train in 1957, in Austin the August before the incident at my grandmother's, and on

December 26, 1985. These experiences usually include more interaction and often take place on the visitor's own turf.

The short visits seem almost always to concern psychological activity, the long ones to involve more physical testing, almost as if preparations are made or results observed during those times.

By the time December 1985 came around, I may have had these encounters at least a dozen times. And yet I never learned from them. Each time the experience took place, I was as frightened, as tormented, as astonished as before.

This is one of the most difficult internal problems connected with the experience. One would have thought that the mind, acting alone, would have compartmentalized all this material together, as it does with recurring dreams and nightmares, so that when I entered the state I would have had reference to other experiences of it, even though—as in recurring nightmares—the material would still have been terrifying.

My actual condition almost seems to suggest that there was an attempt to render me as helpless as possible, by placing me in a state where each experience was perceived in and of itself, without reference to past encounters. Thus each time the surprise was total.

Running through my memories there is a consistent flavor of intense terror. But is it only my terror, the terror of the body, biological terror?

There may be things about contact between beings formed in different biospheres that we do not understand at all. Perhaps they feel some instinctive emotion, too. I have the impression that these experiences are very intense for them, if not actually frightening.

If the terror is an unavoidable side effect of our biology, then the amnesia can be seen not only as an act of self-protection but as one of kindness also.

Assuming the correctness of my perceptions, this book then becomes a chronicle not only of my discovery of a visitor's presence in the world but also one of how I have learned to fear them less.

I look out my window. It is a warm afternoon, cloudy

and thunderous, an afternoon of early spring. People go back and forth beneath umbrellas, their feet splashing in puddles. A helicopter sails across the sky, a jet angles toward La Guardia.

It's all so normal, so home. But what else was that in the sky—a flash of silver light, or something reflecting on my glasses?

The Image

The morning after the hypnotic session covering my experience with the fogbank (March 11, 1986), I awoke not only feeling as if I had been beaten up during the night, I was aware of something new in my mind.

At first the exact nature of this new manifestation was not clear to me. I was oppressed by it; there was an acute impression of being watched. Then I began to realize why: I *was* being watched—there was a face staring directly at me, the grave, implacable, subtly humorous face I had come to recognize from hypnosis.

A vivid image of her had emerged in my mind. It was so real I could almost touch it. This was disturbing and I was eager to expel it, assuming it to be a side effect of the hypnotic session, occurring because Dr. Klein had asked for so many details about her appearance.

It was so extraordinarily clear. I was in a panic. I couldn't live with this image perpetually reminding me of the visitors' enigmatic presence in my life.

I went into my office and sat on the floor, going deep into a state of meditation. I drew my concentration to my body, directing my attention to my physical center of gravity just below the navel, and away from my racing mind.

It took only a moment for me to see that the image had not gone away. On the contrary, it had become far more clear. It wasn't anything like any other imagistic material I had ever had in my mind. I could not calm myself. I was frantic. For the first few hours it was static, simply staring back at me with those large, glistening eyes.

I have never had eidetic, or photographic, memory, so this image was something very new for me. An eidetic image is very much like a photograph inside the mind. This one, though, was far more than a photograph. It had the urgency of life about it.

Despite my attempts to explain it to Budd Hopkins and Don Klein, I could not succeed in communicating to others just how special it seemed to me to be. Nor did I really know this myself until a few days later, when some of the remarkable properties of the image were revealed.

I think that the image was somehow triggered by hypnosis. Maybe the intense state of concentration evoked it from my unconscious . . . or maybe I attracted the visitors attention and they responded.

After the image appeared I did research into eidetic memory and found that it is very rare in adults, almost to the point of being nonexistent in Western cultures. What's more, the descriptions I found of eidetic images did not even begin to correspond to what I was experiencing. People did not report that their eidetic images had a life of their own.

This one seemed ready to reach out and touch me. I felt a strong sense of relationship. Looking at it was more like looking at a person behind soundproof glass than looking at a picture.

I found that the image not only moved about of its own accord, it would move on command. It showed me its hands, its face, every detail of its body. Anne asked me to describe its feet and it leaned forward against something that I could not see and raised a foot, which appeared almost like a very simple version of a human foot. Instead of toes there was a solid structure split in only one place. Like all the joints, the ankle appeared simple in structure.

While I might indeed have been viewing the result of some extraordinary connection between myself and a real, conscious being, it may also be that this was an act of the imagination—the act of a mind calling upon itself to provide another argument in favor of this being an experience with an external component.

If what I was really dealing with amounted to some sort of deep and instinctive attempt to create a new deity for myself, to remain agnostic was to put the conscious me in the interesting position of opposing my own unconscious aim.

What if my unconscious got mad at me and started throwing off things that were really scary, even dangerous? We don't know a thing about conjuring and magic. We've dismissed it all, we who love science too much. It could be that very real physical entities can emerge out of the unconscious. That was certainly one of the hypotheses suggested by what had already happened. I worried that I might not be in control of this conjuring ability at all. I'd already conjured something awfully disturbing. What if there were even more disturbing things waiting in the pantheon of the subconscious?

That was on the one hand. On the other hand, maybe I could make this thing become a real, solid being. Frightening, but also fascinating.

Budd Hopkins suggested that I get an artist to render the image. We chose Ted Jacobs, because he is skilled in creating portraits from verbal descriptions.

It was when Ted came over with his sketch pad that I discovered what was most interesting about the image. I was sitting with my eyes closed, describing this face as carefully as I could. I could see it in amazing detail, moving closer and then farther back, observing fine points such as the faint dusting of white, powdery fuzz that seemed to cover its cheeks and forehead, making it feel, I would imagine, to the touch as smooth as the downy head of a baby. The nose was not very prominent, but the end seemed sensitive, almost like the end of a finger.

As I watched, the image moved its nose, revealing that this was obviously a sensitive organ both of touch and smell. The mouth was not straight, but rather one of those rich and complex lines that come to a human mouth with the advance of years. Centered in this mouth was a remarkable expression, the outcome, it seemed to me, of implacable will leavened by what I can only describe as mirth. Ted Jacobs tried especially hard to capture that elusive quality, and succeeded brilliantly—although the final result, on the cover of this book, is a bit more human than was actually the case. Specifically, the mouth was nothing more than a line, albeit a complex one. There were no lips at all. And the cranium was a good bit larger than the cover portrait would suggest.

The chin was strong, very pointed, and there was a general impression that the skin was stretched over a plated bone structure.

By far the most arresting feature in this face was the eyes. They were far larger than our own eyes. In them I once or twice glimpsed a suggestion of black iris and pupil, but it was no more than a suggestion, as if there were optic structures of some kind floating behind those wells of darkness.

It was those eyes that I saw staring down at me on October 4, those eyes that gleamed so furiously in the faint night light. I remember them from December 26, too, and from the summer of 1957, and from the experience with the fogbank.

Ted asked me many questions about the eyes. When he asked me how they looked closed, I got another shock: The image closed its eyes. I saw the huge, glassy structures recede and loosen, becoming wrinkled, and the lids come down and up at the same time, to close just below the middle of the eyeball.

I described this to Ted, but he wanted to know more. How about a profile view? Had I ever seen a profile? As I sat there staring into the darkness of my own mind, I saw the image obediently turn its head.

I could hardly believe what I was observing. Was this a phantom? What was it? My research thus far has not uncovered any specific paradigm of this experience. I will not assert finally that it was a mental phenomenon as yet unidentified, but at the moment this remains a distinct possibility.

While the image stayed with me, it remained exactly the same as it was when I first saw it. I could observe any part of the body from the top of the head to the tip of the foot. I could do this again and again, and see the same thing each time. On March 13 I made a complete physical description on tape. On March 23 I repeated the description again, then compared the two tapes. There was no difference. The image was unchanged.

Beyond the face, I was able to see the figure's back, the sides of its head, its arms and hands, its feet, torso, abdomen—every part of its body. Under close scrutiny, its surface was smooth but did not seem to have a layer of fat under the skin, which was stretched tight over the bones. The structure of the knee and elbow joints reminded me of the knees of grasshoppers or crickets. The hands were very long and tapered when in repose, with three fingers and an opposable thumb. When pressed down, the hands became flat, suggesting that they were more pliable than our hands. On the fingers were short, dark nails of a more clawlike appearance than ours.

Overall, this did not appear to me to be a highly developed body, but rather a very simple one. There was a general lack of complexity that suggested few bones and not much flesh.

I do not know how to explain this image. If it was not created by the powerful effect of Don's asking me to visualize the creature, then perhaps it was some sort of sophisticated holographic projection. It might be possible to maintain an image in the mind if one knew how to stimulate the optic center in the right way.

Is that what happened? Subsequent events suggested that the image was something even more extraordinary than it at first seemed.

The Visitation of March 15, 1986

Late on the night of March 14, after I had come back from the hypnosis session covering events in our apartment on East Seventy-fifth Street, I sat down once again to think things through.

The image was with me, of course. I wondered what would happen if I asked it to come to me.

Humanity has a long history of conjuring and magic. I have no doubt in my mind that most of this arose from the attempts of helpless people to affect an environment before which they were, in fact, powerless.

But what if that was not the whole story. I sat there looking at it. It looked back at me. Nothing more happened. The thought flashed through my mind almost unbidden that anything I wrote about this experience would be far more intense if I was given some sort of confirmation. It was a true thought: That was exactly how I felt at that moment. The image responded to me with a sharper stare.

On Saturday morning we went to the country. Our son had invited a friend, and we picked this child up on the way. She was one of his school friends, also seven, and the two of them were full of excitement about their weekend together. At no time were the subjects of flying disks, visitors, or any related material discussed at all, and I doubt very much if such things were in the pantheon of either child's awareness. Our son had not been exposed to any of this material and remained totally ignorant of it.

Before dinner I took a walk along our quiet, private road. It was a moderately clear night, with a quarter moon. On the walk I saw a hair-thin streak of light come straight down out of the sky. I thought: I'm disappointed in myself—or in them. Why such a dismal little manifestation?

It was dark when the four of us sat down to the dinner table. We had been eating for only a few minutes when our son's guest suddenly shouted, "A little airplane covered with lights just flew through the front yard!"

There was real shock in this kid's face. The child looked at me, obviously distressed. My impulse was to hide under the table, but I pulled myself together and managed instead to speak in an offhand and reassuring manner. "There's an air base near here," I said. The National Guard base is thirty miles away, but it was all I could think to say. "We don't let those things bother us. Best to just forget about it."

I got up and went outside, but saw nothing. Soon the shock subsided and the children went on eating. Anne and I just sat looking at one another. She had only been hypnotized for the first time the afternoon of the previous day, and knew almost nothing about what was happening with me. From her own hypnosis she had concluded that some sort of visitor experience might be involved, and thus the little girl's statement scared her.

After dinner the two of us went upstairs and discussed the matter. Frankly, the kid's observation, coming as it did at that moment, had convinced me that on some level what was happening must be real. Why else would the child have made that announcement? Not a word about the visitors had been said within earshot of either of the kids, and the little girl was absolutely without information about this subject.

I told Anne about my attempt at communication. "I had a feeling you'd do something like that," she said. "Too bad I can't drive; I'd take the kids back to the city and leave you here to face the music." She stopped. "No I wouldn't." We sat hand in hand in the dark while downstairs the kids read quietly together.

I wasn't sure I could drive the car even if I had wanted to. I could barely keep my eyes open. I recognized the floating sensation of a light hypnotic trance. Was I hypnotizing myself? It's possible.

But what had our son's friend seen? The next day I asked her if she knew what a flying saucer was. She replied, "A what?"

"You know, a flying saucer."

The child looked at me like I was crazy. "I don't know what that is. Your son and I are going out to his clubhouse." Her confusion revealed her lack of knowledge.

When we went back to the city I engaged the child's father in conversation. "Do you remember flying saucers?"

"Wha—yeah."

"Ever read any books about them?"

"Can't say that I have."

"Ever discuss them?"

"Whit, what is this about?"

"Ever discuss them?"

"No. Now what? Do I win or lose?"

A child like I was, brought up in the fifties, would have known about flying saucers. They were big news in those days. But they aren't now, so it's not surprising that the little girl was uninformed, but it is important. It is very important, obviously. The girl saw what she saw, in a simple and real way. When people dismiss such innocent and uninformed testimony, they make a great mistake. Precisely because it is so uninformed, it is powerful evidence of the reality of the phenomenon.

But what reality? Maybe the child really saw an object in the physical world. But maybe, also, the mind has powers that we do not understand. Perhaps there is such a thing as mental telepathy, and when I asked the image to help me, what I really did was send my own inner self on a quest. And at the end of its quest it found this innocent, open little mind, entered it, and there created a hallucination, knowing full well that the little guest would be the last person at the house likely to see anything—and thus the first one to be believed.

By nine both children were fast asleep. I was in a surprisingly benign mood, listening to music on WAMC out of Albany and enjoying being with my wife. We sat together in the parlor in our big upstairs bedroom and got sleepier and sleepier. By the time the clock rang ten it was all we could do to crawl into bed. We went to sleep.

Sometime during the night I was awakened abruptly by a jab on my shoulder. I came to full consciousness instantly.

There were three small people standing beside the bed, their outlines clearly visible in the glow of the burglar-alarm panel. They were wearing blue coveralls and standing absolutely still.

They were familiar figures, not the fierce, huge-eyed feminine being I have described before, but rather the more dwarflike ones, stocky and solidly built, with gray, humanoid faces and glittering, deep-set eyes. They were the ones I felt were "the good army" when they took me on December 26.

I thought to myself, *My God, I'm completely conscious and they're just standing there.* I thought that I could turn on the light, perhaps even get out of bed. Then I tried to move my hand, thinking to flip the switch on my bedside lamp and see the time.

I can only describe the sensation I felt when I tried to move as like pushing my arm through electrified tar. It took every ounce of attention I possessed to get any movement at all. I marshaled my will and brought my attention into the sharpest possible focus. Simply moving my arm did not work. I had to order the movement, to labor at it. All the while they stood there.

I struggled, bit by bit clawing closer and closer to that lamp. I turned my head, fighting a pressure that felt as if a sheath of lead had been draped over me, and saw the light switch in the dark. I watched my hand move slowly closer, and finally felt the switch under my finger. I clicked it. Nothing. Tried again. Still nothing.

The electricity was off. The burglar alarm was still working because it had battery backup—but apparently it meant little to them, as they had entered the house without tripping it.

When I turned my head back I confronted a sight so weird, I thought afterward that I did not know how to write about it. I still don't, so I am just going to plunge ahead.

Beside my bed and perhaps two feet from my face, close enough to see it plainly without my glasses, was a version of the thin ones, the type I have called "her." It was not quite right, though. Its eyes were like big, black buttons, round rather than slanted. It appeared to be wearing an inept card-board imitation of a blue double-breasted suit, complete with a white triangle of handkerchief sticking out of the pocket.

I was overcome at this point by terror so fierce and physical that it seemed more biological than psychological. My blood and bones and muscles were much more afraid than my mind. My skin began tingling, my hair felt like it was getting a static charge. The sense of their presence in the room was so unimaginably powerful, and so *strange*. I tried to wake up Anne but my mouth wouldn't open. The moment I thought of the kids a clear picture flashed in my head of the two of them sleeping peacefully.

The thing before me seemed like a sort of interrogatory. Why the suit? Did it mean that they were showing me a male? If this was a hive species, there might well be more than one sex, and they might be physically very different. Females, males, and stocky little drones?

Now what was I going to do, having called them—lie here and quake? I had wanted to communicate.

They were obviously waiting for me to do something. I saw their faces so clearly, their eyes dark, glittering pits in their dun skin. I could not help noticing that there was a sort of jollity about these beings. I'd thought before that they seemed happy. Perhaps whatever they were trying to do was going well.

They had responded to my summons. What on earth should I say? I wanted them to know that I was still in pos-session of myself, that despite what I can only describe as a terrific assault against me, physically and mentally, I was still functional and on some level independent. More than this, I wanted them to know how I felt about them, despite all the complex connotations of what they were now doing to me. There may very well be good reasons for their be-havior. Have all of their contacts with human beings been peaceful? And how about me: Had I fought in the past?

If they had a hive mind, it might be that the amount of volition I had left was all they could allow me without risk-ing loss of control of the situation. What if I'd been able to do something unexpected very quickly, like reach out and take one of them by the shoulders? Would the hive then have become confused about where this being was? Would it have been that simple to take a captive?

There was and is no way that I would ever make a provocative gesture in their presence. In fact, I wouldn't move at all unless bidden, not until I understand more. If one could escape into their world, one could also get lost in it.

Lying in that bed, I felt a very strong sense of responsibility. I had to communicate in some nonthreatening manner. I was an emissary of sorts—although perhaps only to the court of nightmare. If so it was a strange sort of bad dream, in that the terror began to pass even though the dream hadn't ended.

Again it took an absolute concentration of will, a centering of my attention and the application of the most careful effort to the muscles of my face, but I did manage to smile.

Instantly everything changed. They dashed away with a whoosh and I was plunged almost at once back into sleep. Now I did dream—qualitatively a very different experience from what had just transpired. Frankly, I'm quite certain that the beings I saw were not a dream, and probably not a hallucination. What they were was an enigma.

Interestingly, my dream was an unfrightening repeat of one of the few really terrifying nightmares I have ever had. This was of being chased through a stark stone palace by a robot with beady pop eyes. This time, however, I didn't run and the robot finally sat down and contented itself with staring at me.

The next thing I knew, morning came. I opened my eyes, feeling absolutely drained. Anne said, "Well, it was a quiet night," and proceeded to make a beautiful breakfast while I sat and stared.

Everybody was happy and well around the breakfast table. The *Times* was as thick as ever and the coffee and waffles were delicious. I was back in my world again, with my own familiar family. When I told Anne about what had happened, she laughed merrily at the idea of the painted suit of aged design, and reset the clock on the stove, which had lost five minutes during the night.

I found that one of the other people I have met has also had an experience involving visitors in archaic suits. This suggests that the visitors are not too interested in our clothing, or confused about its significance . . . or perhaps that their thought processes have not gotten very far yet in regard to clothing. It may be that, if we ever meet them openly, they will not be quite as naked as the creature who emerged in *Close Encounters*. Perhaps they will be wearing double-breasted suits circa 1952, complete with pocket handkerchiefs.

What happened the night of March 15 was fundamentally different, and more open, than any other contact I have had. The visitors almost irrefutably announced themselves to me. They allowed me to see them while in full possession of all my other memories of them—albeit in a more or less completely restrained physical condition. And they preceded their appearance to me by the witness of the uninvolved child, the one person there that night who had absolutely no relationship to this at all.

Who had come to see me during the night? Did they really drop down from the sky, or have they come from some other cosmos, a place where dreams are real and reality a dream, where shadows and those who cast them are one and the same?

Weeks after writing the material preceeding in this chapter I met a woman who said she thought of her visitor as a man, and proceeded to describe a small being, very gentle, with round eyes like shiny black buttons and a tiny, almost nonexistent mouth.

This was what I had seen, or a model of it. The suit must have been a form of communication.

Why not simply speak? They have a voice of sorts. I have heard it, others have heard it. And they can also speak in the center of the head. I wondered if it all had something to do with my request for confirmation. Were the three real ones holding a scarecrow near me because they wanted to see what

I would do with my limited physical mobility and did not want to expose a living being to the danger of my touch?

In retrospect, I am glad that I did not reach out. My impression is that these people, if they exist, are more than a little afraid of us: They are deeply afraid. I suppose it was best to smile rather than move my hands toward them, but I wish that there had been touch. Could there have been, or would my fingers have crossed only air? I suppose that I will always wonder.

I asked for confirmation, not proof. It seems that they took me at the exact meaning of the word.

FIVE

*I rage to know
what beings like me, stymied by death
and leached by wonder, hug those campfires
night allows,
aching to know the fate of us all,
wallflowers in a waltz of stars.*

—DIANE ACKERMAN
"Lady Faustus"

ALLIANCE OF THE LOST

Recollections of My Family

Trapped in the Dark

From the beginning, I had been disturbed that my wife and son might have been involved in this. At the least, they had suffered with me through my upheavals. At the worst, they were as entangled as I am. Our son has been preserved from almost all conversation about it and from direct experience of my personal trauma. Even before I knew exactly what was happening to me, my first concern was to leave his happy childhood intact.

When I realized that my memories fitted in with material described in *Science and the UFOs,* I did not tell Anne. I didn't tell anybody. As I have said, my initial impulse was to hide everything, and when I could not do that I sought professional help through Budd Hopkins. I told Anne nothing and our son less than nothing.

Thus when Anne was first hypnotized to aid her in recalling the nights of October 4 and December 26, she was aware only that something very unusual seemed to have happened on those dates. I did not keep it all from her out of some desire to preserve her the purity of hypnotic recall. I kept it from her because there was a possibility, however remote, that visitors had been around us in the night.

Thus the "visitor hypothesis" had been discussed by us only in general terms. Only after her first hypnosis, which took place on March 13, 1986, did I indicate that I felt there might have been some nonhuman presence involved in our lives. On the evening of March 15, she was treated to the spectacular witness of the little girl at our country house. It is a testament to Anne's courage that she remained there that night.

It is also true, though, that Anne has a very active mind, and she was far too curious to simply retreat. Once she understood that something might be happening, our familiar teamwork system began to function, and she took over the intellectual direction of our study, bringing to it her own creative and open mind, as well as her steadfast insistence that all speculation proceed from known facts.

Before hypnosis Anne had a vague memory of me warning her of fire on the night of October 4, of hearing an explosion and our son calling out to me. Neither of us can understand why she did not respond to the warning, in view of the explosion. She had no memories at all of the night of December 26.

There were also some prior memories. The previous March I had told her that I "flew around the room" one night. I remembered this as a vivid dream. But flying dreams aren't unusual, not even faintly pathological. They are generally associated by psychiatrists with hidden desires to escape stress. In 1982 we had the experience of the white thing, which was dealt with in detail in Anne's hypnosis.

On March 13, 1986, Anne was hypnotized by Dr. Robert Naiman. We chose a psychiatrist other than Don Klein so that there could be no possibility of his questions taking on some sort of unnoticed direction because of what he already knew.

I felt that there was a real chance here to find some answers, and I wanted to do everything possible to encourage that outcome. Anne and I are a very deep, total marriage. If something was *really* happening to me, then she had to know. She would have some involvement. If she had reported nothing, then to me it would have indicated that mine was an essentially psychological experience—perhaps shared in some unusual ways, but essentially psychological. Thus the visitor hypothesis would no longer have been among the more likely ones. I think that I would then have concluded that a hitherto undiscovered mental process with a definite physical effect was probably operating.

Bob Naiman had worked before with people who have had this experience, and took the same healthy and supportive stance that Don Klein did.

Budd Hopkins was present at Anne's session. Questions asked by him are identified with his name. All other questions were asked by Dr. Naiman.

Despite all the progress I had made in dealing with this experience, I must admit that Anne's hypnosis disturbed me all over again. It was clear that she was not specifically expecting or seeking anything. And yet *something* was there, and in a way that subtly implied that it has had a profound effect not only on her memories but even on her role in our life together.

Her hypnosis does not reveal a person trying to concoct a story, but rather one trying hard to avoid remembering something she has been told in the strongest terms to forget. She was compliant all right, but not with the hypnotist. She complied, it appears, with something else that issued previous, stronger suggestions. And they overpowered the hypnotist's efforts for a very obvious reason. My wife appears to have been made to believe that my mental health depends on her not remembering, on her providing me with a safe haven in ordinary reality when I need one.

I suspect that she is right to believe this. Even under hypnosis she protected this role, which is probably essential not only to my mental health but to that of the whole family.

The regression began with her memories of the night of July 30, 1985. She did not know it at the time, but there was some evidence in one of our son's school journals that an event involving her might have taken place on that night, when she and he were in the country and I was away on business. Rather than suggest to her in any way that we thought this might be the case, Dr. Naiman began with that night without telling her why.

Hypnosis

JULY 30, OCTOBER 4, AND DECEMBER 26, 1985

SESSION DATES: *March 13, March 21, 1986*

SUBJECT: *Anne Strieber*

PSYCHIATRIST: *Robert Naiman, MD*

Dr. Naiman: "First we want to concentrate on July thirtieth, 1985. You were with your son?"

"Yeah."

"You were in the country then?"

"Yeah."

"Who was there?"

"A lot of workmen came that day, so I wouldn't be too lonely because I didn't have the car. Whitley took the car to go to the city. The workmen were going to be there, so I wouldn't be too lonely. I believe I bicycled to the store. I went to the store. I remember thinking how am I going to go to the store. I wouldn't want to drive anyway but I've got my bicycle and I can leave my son because the workmen are there. And I did. We wanted something to make some treat or something. We wanted some snacks or something."

"He wanted some snacks?"

"And I did too. I remember it was going to be lonely putting him to bed that night and it was. I don't remember anything strange."

"Was it unusual for you to be there alone with him?"

"Yes, because we always all drive up together and I can't drive very well, and I wouldn't want to drive so I'm never alone there with just our son overnight."

"I want you to concentrate on coming back from the grocery store with the treats."

"Yeah."

"When was that?"

"It was afternoon. Late afternoon. About four or three, I think, because the workmen were there but they were leaving."

"What are they working on?"

"Building the pool on the deck."

"And they left around four, and it was just you and your son?"

"Yeah."

"And then what?"

"I don't remember what we had for dinner, but it would have been something fairly simple. I might have baked something, but I don't remember. I think I went to get some chocolate chips, and we made cookies. It must have been earlier in the day because I might have given some cookies to the workmen. I think we did. Maybe that was another time, but it could have been that time. I remember we did that once. And that's the kind of thing I would have gone to get. I wouldn't have had any chocolate chips, and I think that's what I went to get. I wanted to get a paper and I wanted—yes, I remember that. And I said, 'Why should I wait when I have a bicycle?'"

"So you and your son had dinner together, the two of you?"

"Yeah."

"What time would that be?"

"Six or so. [Sounds perplexed.] I don't remember dinner. Were we invited somewhere? I don't think so."

"And what time would you be putting him to bed?"

"Around eight. Seven-thirty."

"Something you were not looking forward to?"

"Well, it's hard being all alone all day with a kid and I'm tired at night and I'm not usually the one who puts him to bed. And I'm not as good at reading stories as Whitley is, and I don't look forward when Whitley's gone to putting him to bed."

"But it went all right?"

"Yeah . . . I don't *remember*. No . . . I couldn't have

watched TV because we had no TV reception. But we had the VCR. I don't think I had a movie to watch. I don't remember. I remember Whitley came back and it was earlier than I thought."

"When was that?"

"The next day."

"Do you remember anything that night, when you were sleeping alone?"

"No—I—well, no."

"Did your son call you during the night?"

"I don't think so."

"You slept your usual sound sleep?"

"I think so. But it's lonely going to sleep in the country alone at night. I might have heard some noises. It seems like maybe I did, but they weren't anything because I had the doors locked."

"And you have a burglar-alarm system?"

"Yeah."

"Do you often hear steps . . . sounds?"

"Not steps . . . I doubt if it was steps. You don't always hear sounds. It was pretty quiet. Not noisy."

"All right, so let me give you one more minute to concentrate. I want you to concentrate as hard as you can because you have this very special brand of concentration ability right now, this capacity to concentrate. I want you to just concentrate on that night from sundown on."

"Funny. I remember the afternoon but I don't remember the night. I don't remember after the workmen left. [Long pause.]"

"All right, we won't put any more time in on that right now. But it's very possible that after I take you out of the trance, between now and the end of the weekend something will come to you about the thirtieth. And if it does, try very hard to remember. If something passes through your mind."

"Yes." (Nothing ever did. She was left with a memory that just stops right before dinner and doesn't start again until I returned the next morning. What happened in between was completely blank, as if powerfully blocked. Her earlier memories of that day are perfectly normal.)

"Because of this procedure we are going through now, we may be loosening up some memories that will not emerge until you're out of the trance. So be alert."

"Yes."

"Now let's go to the night of October the fourth. As I understand it, you and Whitley were there and your son was there and you had guests, Jacques and his woman friend."

"Annie."

"Everybody's going to bed. You've had a good time, a good meal, and a lot of good wine and conversation. Is that true?"

"Well, we went to a restaurant."

"Oh."

"Next day we had fun because Jacques went swimming, and the water was very cold."

"Let's go back to that night of October fourth. You've said good night to your guests and your son is already asleep, of course—"

"We got home late. We didn't get finished at the restaurant until about nine. They had been there before but they slept on the couch. This was the first time they saw the guest room. It didn't have a bed in it before. So dark you couldn't see anything very well though. Everybody just came home and got our beds ready and put on our pajamas. Because we were all tired. I don't think we talked much that night."

"Yes?"

"We left later than usual or—I don't remember. I think that's why we ate out. There was no time to buy groceries. So we must have left later than usual, for some reason."

"This was a Friday evening?"

"Yeah."

"So you said good night to everyone. And you and Whitley went upstairs?"

"Yeah."

'Just give me to the best of your recollection what happened that night, concentrating as hard as you can."

"[Long pause.] It wasn't a peaceful night, but I don't remember why. [Pause. Seems distressed.]"

"What are you thinking of right now?"

"Well, I don't know."

"Because you just screwed up your face and clenched your eyes."

"It seems like there was a lot going on, but I don't remember. I—I—remember when Whitley thought the roof was on fire. I don't remember that. But I remember it was like a culmination of a lot of other—it was like—it was surprising, because it was like a culmination of a lot of other activity. I don't—I don't—seems like it was late and not dark but I don't remember that, and it's not clear. But it doesn't seem like it was dark enough. Usually it's so dark. It's all *dark*. And so restful and quiet, but I don't get that feeling about it. I get the feeling that Whitley was up all night, and it was this thing and that thing and finally it was the roof. It was something else other than the roof on fire. There was something about the stove for another reason."

"What's going through your mind now?"

"Nothing."

"Nothing? Concentrate hard."

"I just see a light. I mean, it's not dark. You know, it's not dark."

"Yet when you came back from the restaurant you were struck by how dark it was."

"Did I say that?"

"Yes."

"I remember the house was dark, because we couldn't see. I remember thinking they can't see the guest room very well and they'd never seen it before. Of course they could have turned on the light. It was very dark outside. I don't believe we turned on many lights. We just went to bed. We were all very tired and we all just wanted to go to bed. The thought flitted through my mind, you know, as the hostess, should we drink something or sit and talk, but we all just wanted to go to bed."

"What time was it?"

"I think it was about nine."

"When you came back from the restaurant?"

"You were very tired. You get tired earlier out there. It's funny. Restful. Affects everybody. Never stay up late."

"I don't know what the calendar says, but I get the sense that there was a new moon that night. Very little moonlight." (Dr. Naiman had not been told that there was a heavy fog all night and it was inky black due to the lack of reflected light in this sparsely populated area. The moon was waning and past the half. It rose about 10:30 and set in the predawn hours.)

"No."

"No what?"

"I—it—I don't know. It seems like there was a light."

"Tell me about it."

"No. I don't think there was. I think I just think there was. I mean—it's just that I have my eyes closed and it's not dark. It's light."

"You're back on that night, October fourth?"

"Well, I'm trying to be. I'm there and we go to bed. I remember when we're in the restaurant more clearly. Then we walk to the parking lot and the car's cold. Very cold. It's dark, but the restaurant's illuminated outside. It does seem to be a dark night, yes it does. But the restaurant has lights aiming at it, so maybe the contrast . . . but it seems like a very black and inky night. But a clear night? Stars? I don't think it was gray and cloudy because when it's cloudy it seems light. But when it's clear you can see the dark night and you can see the stars. But I don't—"

"Let's go back to your being in bed. It was sometime after nine o'clock that night."

"It was odd being in bed with people in the house because you feel like you can't talk loudly. Our house is not very soundproof."

"What?"

"Our house isn't very soundproof."

"So you had to be quiet?"

"You feel like if you bounce around in bed they can hear

you and they can hear you talk so you kind of whisper and feel self-conscious. And it was odd to think of so many people in the house because usually there's just my son down there and it's very empty . . . in the kitchen . . . you feel like it's empty and it was so full. The house was *full,* you know."

"Did that give you a feeling of security?"

"No. It was just different."

"But you do remember lying in bed with Whit and whispering?"

"Vaguely. Not very clearly. I don't remember much about that."

"Were you comfortable in bed?"

"Well, we were kind of—yeah, we're always comfortable in bed."

"On a night in October, it must be pretty chilly up there."

"October? No, November. It's December."

"No, October fourth we're talking about."

"October?"

"It was October that Jacques and Annie—"

"So it was. I think it was December. Because I remember snow. But it didn't snow in October."

"It didn't?"

"It's not likely. I remember snow." (Either she has confused the October and December experiences, or she has a vague memory of the chilly fog.) "I remember it was very cold."

"Do you remember the kind of clothes you were wearing when you went to the restaurant?"

"No. But they'd be casual clothes. Maybe they'd be my daytime clothes. A skirt . . . I might not have changed clothes."

"I'm interested in exploring how cold you were. Were you wearing enough clothes?"

"I might not have been, because I often leave my country clothes up there. It's always cold in that house, because you have to get the fire up. I would have been cold, but I turned on the electric mattress pad and got warm."

"How did your body feel that night?"

"Well, I think it was cold and got warm. Our room was very warm when Whitley woke me up."

"In the middle of the night or the morning?"

"Oh, it was during the middle of the night. Yeah."

"Tell me about that."

"Whitley'd been talking about the chimney lately. Feeling it. And he thought the roof was on fire. But I didn't see how the roof could be on fire because there were no flames and there was no light." (I woke her up the first time when I was awakened by the light that passed by the windows. By the time she was aroused, it had reduced to a small glow in the front yard.) "If the roof's on fire you'd see the roof all lit up. He saw a light that I didn't see."

"Did he tell you about the light when he woke up?"

"Well, he said . . . I don't remember how he did but I had the impression that he saw flames or light. Not flames. It didn't make sense to me."

"Is it possible you didn't open your eyes?"

"Yeah."

"That's possible?"

"Yeah."

"Yet you do make some references to a light that night."

"That feeling is gone now. But I don't remember it as a restful night."

"You know, Budd is here of course. And you don't mind if he asks some questions?"

"No."

Budd Hopkins: "Did you have any dreams that night?"

"Don't remember."

Budd Hopkins: "You said it had been a restless night. Because of dreams?"

"Let me think. I don't think Whitley was there very much. He was gone. You know, he goes sometimes at night. He goes and works. Or he just goes."

"Where did he go that night?"

"Downstairs."

"You have an impression of Whit being away from the bed?"

"Yes. It's lonely, you know. I wish he wouldn't do that."

"Was it after he said the roof's on fire?"

"I think it was before too. He went out, then he came back again. He just was doing things all night."

"It certainly was not a night of sound, deep sleep for you, was it?"

"Well, it doesn't seem to be but I don't remember anything. But it has to be really."

"Did you hear your son?"

"Yes!"

"You heard that?"

"Oh, yes, he sounded so frightened. Really scared."

"Is that very common?"

"He gets nightmares sometimes. But he sounded especially frightened. I remember he sounded really terrified. So frightened. More frightened than usual."

"He screamed, eh?"

"Oh, yes. *Oh, yes!* It's painful to hear." (Nobody else remembered him screaming, only calling for me.)

"Is that something that in a normal night's sleep you might sleep through without hearing?"

"Oh, no, no! Whitley usually hears him first, but I always hear him."

"You didn't sleep that soundly?"

"Oh, no. I heard it."

"I know you heard it. But I want to know if you heard it—"

"Oh, no. I heard it. Some nights I might not, but usually I hear it. This you couldn't miss. I mean, it was so loud."

"Did he say words?"

"Well, he did but I don't remember what they were. He was really scared. Something *really* scared him. I thought maybe something was happening to him, because it was like something was happening to him. I thought somebody was doing something to him. It was a different kind of scream."

"Why didn't you go to him?"

"Because Whitley was already on his way. But I remember feeling very uneasy. I wanted to go too, but I felt I shouldn't."

"Why not?"

"I thought there was something Whitley would— It had to do with him. He was *supposed* to go."

"He was supposed to go?"

"He was supposed to go. I wasn't supposed to, but I wanted to go."

"It must have been difficult staying in bed."

"It was, because I wondered what had happened."

"When did you find out what had happened?"

"I don't remember. I don't remember."

"What did happen?"

"I *do* remember he was gone for a long time. He didn't come back. Sometimes he goes and sleeps down there when our son has a nightmare. Or sleeps in our son's room in the apartment. I remember feeling very lonely that he didn't come back, and I didn't think it was fair to be left like that. It was very lonely and scary. It made me uneasy. He kept going, you know. He kept *going*. (I never sleep with our son.)"

Budd Hopkins: "When your son kept crying?"

"Whitley kept going. Kept leaving."

Budd Hopkins: "I want you to do something. You're lying very, very quiet. Relaxed, just as you were that night. I want you to concentrate on what you can see and feel and hear. If you can see something through your eyelids. Feel your body, your shoulder, your legs. Feel relaxed."

"I don't feel relaxed. I'm not relaxed. I can't feel relaxed if I wasn't relaxed. I mean, it wasn't relaxing. No, it wasn't relaxing at all. There was too much going on, you know. It was—"

Budd Hopkins: "Was there something going on in the room, or did Jacques and Annie—"

"No, they weren't in on it. There was something going on. I wanted to know what was going *on*. It looked—things were going on and I wanted to know what was going on!

There was lots of things going on and I couldn't figure out what was going on!"

"Why didn't you get up and go see?"

"I couldn't, because I wouldn't. I— Was I afraid to, or wasn't supposed to? I wasn't supposed to. It was like your mother said to you, 'You have to stay here,' even if you don't—you're dying to get out and see what's going on, but you know because you've been *told*." (Her identification of the directing force as feminine is fascinating.)

"You were trained to do that?"

"Well, we're all trained to do that from childhood."

Budd Hopkins: "Who told you that?"

"Nobody told me! I just had to do it."

"Is this something Whitley told you?"

"No. He's just left. No."

"Have an impulse to turn on a light?"

"Oh, no. No. I wasn't supposed to see."

"Who said so?"

"No one said so. I just knew it."

"You weren't supposed to *see*?"

"No, and I just knew it. That's what worried me, because I wasn't supposed to know but my son was so *afraid*. And Whitley was saying things like 'The roof is on fire,' and I wasn't supposed to do anything. It's like somebody says, 'Well, the car is crashing but don't do anything'!"

"Strange orders."

"Well, they weren't orders. You see, they *weren't*. They weren't orders, no."

Budd Hopkins: "Anne, I want you to do me a favor. I want you to—with your eyes closed and very relaxed—"

"I'm not relaxed."

"As relaxed as you can be. I want you to have a little dream. A fantasy. About what all that activity was. What's happening?"

"All right."

"Somehow Whitley's involved. Your boy is involved."

"I'm not involved!"

"Well, you'll dream about it. Tell us about what you remember."

"Whitley's supposed to go. They came for Whitley."

"I'm sorry?"

"They came for Whitley and he's supposed to go. But I'm not supposed to go."

"Who came?"

"Nobody that I know of. He just has a feeling that he's supposed to. And it's like when someone's going off to war or something, they're supposed to go and you're supposed to stay home."

"Now, there has been a shift, though. Because in the early part of the evening when you thought Whitley was out of bed—you say he often goes downstairs and writes."

"Not in the country. Because he writes upstairs in the country. He doesn't write in the country very often. You have to turn on the overhead light to write in the country. No. I just mean at home in the apartment, he gets up. I find out that he's done a lot of things at night. Or I kind of sense it."

"In the city?"

"Yeah."

"So his being out of bed that night—"

"No, it wasn't usual for the country, actually. He stays in bed in the country. He really does. And that's why I think he gets more rest there. Because he goes to bed and stays and there's no place to go, there's nothing to write, nothing to tempt him, and I'm the one who gets up early and reads in the country. He even sleeps late."

Budd Hopkins: "Why do you think Jacques and Annie don't get up? They heard screaming from your son. Weren't they concerned?"

"I think they did get up, didn't they?" (No. Neither of them testified that they got up, and both of them were awake enough to remember their own movements and that of the other.)

"Did you hear them?"

"I think I remember hearing . . . Annie . . . speaking to him. I think Annie went to him first. And I remember feeling . . . feeling jealous that I couldn't go. I'm his mother. It wasn't right. It made me look bad. It made me look like I didn't care." (Annie Gottlieb did not leave her room, and did not speak to our son.)

Budd Hopkins: "When you hear him calling, do you feel your legs tensing?"

"Yes! Usually Whitley goes but this one sounded bad and I wanted to go. It sounded different and I wanted to go too. Wha— I thought there was something in there?"

"What do you mean?"

"I don't know. It's like there was a friend or something. It's just a memory." (Afterward, she said that this referred to our bedroom, that she thought she had seen "a friend." She would not elaborate. When questioned again about it two weeks later she had nothing to add at all.)

"What was restraining you?"

"I didn't feel restrained. I just felt like I wasn't supposed to go."

Budd Hopkins: "Have you obeyed other things like that in your life?"

"I wonder if I have."

"Is it a familiar feeling?"

"No . . . no . . . But I used to always do it, you know."

"Do what?"

"If there was choice, I'd do it. Because if you do it at least you've done it, you know."

"What do you mean?"

"I just mean that I don't think I was a person who didn't do things. That's not true."

"So this is a variant."

"It's not though, because Whitley's the one that gets up at night."

"OK, so can we shift now to morning?"

"I don't remember it specifically. I don't. I'm trying to remember. I don't remember what we had for breakfast. I remember going swimming. Wanting to see if Jacques could do it. It was cold. I couldn't do it. Or did I? I don't think I even tried. I put on my bathing suit but I couldn't even get my feet in. I felt bad because Whitley got in. But nobody but Jacques got in. Not even Whitley got in. If he did it was only for a short period. Annie did, I thought just to stay even with Jacques because she's smaller than I am. We all wanted to see if Jacques could; it's kind of a joke."

"You don't remember what you had for breakfast. Do you remember the atmosphere around the table?"

"It was pleasant, I think. Pleasant."

"How was your son?"

"I don't remember. He was fine."

"All right. Is there anything else you want to say about that October fourth night. October fifth morning."

"Well, I found it funny when I woke up that the roof wasn't burned."

"You found it funny that the roof wasn't burned?"

"Yeah. 'Cause I thought it was gonna be."

Budd Hopkins: "What about the bang?"

"Maybe that's why it seemed so active that night."

"What do you mean?"

"Maybe it was noisy."

"What kind of noises?"

"[Long pause.] Don't remember "

"Can you describe them in any way?"

"Just our son. Seems like there were a lot of noises, you know. Doesn't seem like it was a quiet night. I get the impression—it wasn't a quiet night. I get the impression that someone was there but it wasn't Jacques and Annie, because they were in their room and they stayed in their room. But—and then I remember Annie comforting our boy . . . it was a woman. . . . I thought it was Annie. It *was* Annie."

"You recognized her voice?"

"Yeah. I think I did. I think I did. Oh, I just get the impression . . . it's general and vague . . . my memory is just no good. But I get the vague impression that they were in there like a cocoon. Locked in that room."

"Jacques and Annie?"

"Yeah."

"How did that happen?"

"Well, it's just because they were. It's like they couldn't get out, you know."

"You think they were paralyzed or something?"

"Well, I just knew they were in there. And they just weren't going to come out. And it was kind of odd because one of the things about sleeping upstairs is that you think the people who are downstairs might get up and walk around and you might hear them and they would hear you, but I knew that wasn't going to be the case. The feeling is very vague. I just remember feeling that. That they weren't going to come out. They shut the door and they weren't going to come out. I remember the morning, I came down and the door was still shut and I thought, *Oh, they're still in there and I wonder if they can't come out, or if they will come out.* I knew they still had to be in there, but it was almost as if they weren't."

Budd Hopkins: "Do you feel, at this point, that she was talking—"

"She was trying to comfort him."

Budd Hopkins: "So she did come out."

"Yeah, if it was Annie. It must have been Annie."

Budd Hopkins: "Did you hear words?"

"I think I did, but vaguely, like 'What's the matter?' I'm not that sure, though. I just remember it vaguely. But I remember thinking that everybody else got there first."

"I want to jump ahead to December twenty-sixth of 1985. What do you remember of that day, as you concentrate very hard."

"Day after . . . I don't remember anything."

"Think hard. Who was there?"

"Just us."

"The three of you."

"Oh, yeah."

"That was the day of the owl?"

"Well, that's what I'm told. I remember the owl. I remember Whitley talking about a crystal, too."

"About a what?"

"A crystal in the sky. But that was before the owl."

"What does that mean, a crystal in the sky?"

"A bright crystal in the sky."

"Did you see it?"

"Oh, no."

"Why do you say 'Oh, no' as if—?"

"Whitley saw a lot of things that I didn't see at that time."

"Did you look for it?"

"Oh, no. Because I knew it wasn't real."

"How did you know it wasn't real? Whitley's a fairly down-to-earth guy—"

"No, he isn't."

"He's not?"

"No. Because there couldn't be a crystal in the sky. He said it had a point that touched the earth."

"It didn't surprise you hearing Whitley, that he sees things like that?"

"No."

"It's an old story?"

"No, not like that. No."

"Why didn't it surprise you?"

"Well—I guess I just thought he'd explain it later. Whitley, you know, said he'd flown around the room. What do you say to something like that?"

"When was that?"

"Oh, he was saying that last year."

"You think Whitley should go to a psychiatrist?"

"No."

"No?"

"No. Because he—I think he can deal with these problems."

Budd Hopkins: "Back to that night. It was so restless—"

"It was like a party. [Nervous laughter.] There are lots of things going on here now. It was like a party and not being invited."

Budd Hopkins: "A fun party?"

"Oh, no."

"What kind of party was it?"

"Well, you know, Jacques and Annie weren't invited either. It was all going on downstairs. And I had to wait for them to come back." (Because this question about the night of October fourth was asked during hypnotic regression to December twenty-sixth, she is now confused about events on the two dates, and it is not possible to tell whether she means that things were going on downstairs on the fourth or the twenty-sixth.) "It was like your mother says, 'No, you can't go. You have to wait for us to come upstairs.'"

Budd Hopkins: "What I wanted to ask is, do you think that feeling may have ever happened to you before?"

"What feeling?"

"That feeling that there's something going on that you're not allowed to see, some kind of activity like that at night."

"Well, I've often felt that there are things going on with Whitley that I wasn't supposed to know. I'm supposed to kind of help him afterwards to deal with it. That's my role. But I can't stop them, you know. He just has to."

"Do you think these are things that come out of Whitley's head?"

"No, I don't think he has hallucinations, no. But I think they come to him because of his head. He has a very unique head."

Budd Hopkins: "Anne, I'd like to ask you—there was a night on LaGuardia Place."

"On LaGuardia Place, yes."

Budd Hopkins: "That something thumped you."

"Oh, the white thing."

Budd Hopkins: "You have a sense of what that was?"

"Oh, yes!"

Budd Hopkins: "Tell us what that was."

"Like a sharp jab in the stomach, right here. [Points to area just below ribs, in center of abdomen.] And it was like four fingers, not just one finger. It was—*oof!* It was like a joke. But then who would do that? And once and disappear, you know. Woke me right up."

"Did you open your eyes?"

"I don't think so. But I sat up. It woke me up. My son woke up and had a nightmare at the same time and said something poked him in the stomach. And then Whitley—I don't remember when it was, it was the next morning or whatever—and he said something poked him in the stomach. He said he saw a little white thing and our son said he saw a little white thing and the baby-sitter said she saw a little white thing."

Budd Hopkins: "Try to make a guess about what that little white thing would look like."

"A little ghost. A little white ghost with little feet and kind of running around and getting out of your way quickly. When it pokes you, you know. The baby-sitter said she thought it was some boys with sheets over their heads, but it didn't look like that." (On reading this description of "little feet" in our apartment in 1982, I was reminded of Annie Gottlieb's memory of "scampering" in our country house in 1985. At the time of her hypnosis my wife was unaware of Annie's testimony.)

Budd Hopkins: "Didn't look like what?"

"No, it had a kind of square head and it's white. . . . I can't see it, Budd. That's just how I imagine it would look."

Budd Hopkins: "Did it have folds?"

"No. Just kind of glowing. Just so you could see it. Otherwise how could you see it in the dark?"

Budd Hopkins: "Did it have any color?"

"No, white."

Budd Hopkins: "Does it speak to you in some way or another?"

"No."

Budd Hopkins: "What do you think it's doing?"

"I don't know. It seems to be kind of a joke, you know."

Budd Hopkins: "It had arms and legs?"

"Yeah."

Budd Hopkins: "Fingers?"

"Yeah. Don't think it had toes, though. Kind of pointed feet. But it was like it wasn't wearing anything but it was, because you didn't see any seams or clothing or anything, but it wasn't naked, you know. Little pointed feet."

Budd Hopkins: "How tall was it?"

"Oh, about as tall as a four-year-old. It had little pointed feet."

Budd Hopkins: "Do you think he was in there, to your knowledge, more than once?"

"What do you mean?"

Budd Hopkins: "Did you see it more than once?"

"No, I was just poked once. Trying to remember if I ever saw him before. Back in my childhood. You know something. Now wait a minute. [Pause.] I think I did. But I don't remember where. You know, I had a very lonely childhood. I was always alone, but I wasn't, I don't think, really. But I didn't have imaginary friends. I didn't believe in that. I wonder if he's in that room. The room glows. I wasn't afraid of the little white thing."

Budd Hopkins: "When you were on LaGuardia Place?"

"I thought it was interesting that he would actually show himself to the baby-sitter. [Laughs.] I thought that was very mischievous of him."

Budd Hopkins: "Anything threatening about him?"

"No." (She apparently does not remember screaming when she was awakened by the prodding. But I remember it as the only nightmare she has ever had.)

Budd Hopkins: "Frightening?"

"No."

Budd Hopkins: "A cute, lovable, little—"

"Well, not really, no. Because, you know, he's invading your privacy—he should stay away. Mind his own business. Just felt, now that I think of it—I didn't feel it at the time—but now that I think of it, it seems familiar. And I feel like I knew him when I was a kid, you know, because—but I don't remember anything at all. But I don't think that's true. I don't think it's true. It's just now that I think of him I feel a familiarity coming over me except I really don't think it was true. No."

She was then brought out of the trance, after making a comment under hypnosis that she thought her memories had been "taken out," an assertion that the hypnotist assured her would not be true in the future.

Subsequent to this hypnosis Anne felt disturbed that her memory seemed to blank at crucial moments. After hypnosis she did not recall her comment about seeing the light behind her eyelids. When questioned about this, she said she was very unsure about it. Maybe the light she was referring to was simply that in Dr. Naiman's office. She decided to attempt hypnosis again, and a week later was hypnotized by Dr. Naiman, who still had not been informed of the results of my sessions.

Before this hypnosis, Anne was talkative and her memory was excellent. During hypnosis it was found that she was still incapable of remembering much about the crucial nights, except that she had a powerful image that the screaming she had heard was me, not our son. She also saw my face as I screamed, and was terrified by the idea that something could frighten me that much. The elusive female presence that is referred to in the first session acquires a more specific existence this time.

Unfortunately, during discussions about her first session I inadvertently let slip that I thought I had been screaming on the night of October 4. Even though her memory may thus have been tainted, her recollection of this is so vivid in the transcript that it may also be that it really happened.

As most of the session was taken up in a futile effort to dislodge memories that either are not present or cannot emerge, I will transcribe only the relevant part of the material. Prior to hypnosis Anne and Dr. Naiman spoke about her reasons for returning.

"I couldn't remember very many things experientially like you were there. I only remembered one or two things in the whole hour and a half as an experience, the way you think of a real memory . . ."

"Just addressing myself to the tape that is rolling, this is the twenty-first of March, 1986, and Anne is here answering my questions on her reflections about her last visit here, which was a week ago today, and that's what she's been talking about. How about the hypnosis, do you have any feeling about that?"

"Well, I found two things. One is, I don't know how very deep I was into it because I did find it very hard to bring up pictures in my mind. But that's I suppose for other people to determine, or myself to determine with more experience. Number two, the wonderful thing about hypnosis, the reason that it gives you a good feeling, is that the barrier that is always there when you are talking to other people—even about very mundane things about which you have no secrets—is lifted, and you feel that you're being very honest, not in a way that someone's going to make you say something you don't want to say, I didn't get that feeling—but you feel a sense of freedom, that you can really be honest. It's very refreshing because you feel that you just have to think about what you want to answer, not how the other person is going to perceive it."

"How come you're back in here today?"

"I think because this is my project as well as Whitley's, and I don't think—I think we have to try once more before I get let in on things . . . and it was kind of odd, I thought, because the things I did remember were not to me very clear memories at all, and I almost feel that they seized upon them. If I have a faint thought that it looks a little light behind my eyelids, but I don't know if it's because this is a white room or because I'm remembering it was light then . . . it was so vague. . . . I feel kind of bad when people try and be very fair. . . . I just don't think . . ."

"One of your last comments was, 'Now I can go home and hear Whitley's tapes.'"

"But I decided not to. I went home and talked to Whitley about it and at that point we decided not to."

"I see. And at that point you had not yet decided to come back here today."

"Well, I figured if I didn't listen to them I would come back, but I said to Whitley, 'What do you think? I think perhaps I ought to do this again before I listen to the tapes,' and Whitley said yeah."

"But you initiated this?"

"Oh, yes. I'm here voluntarily. If I had said, 'No, I'm through with this,' I wouldn't be here."

"How do you feel about this?"

"It's interesting."

Budd Hopkins: "A basic ground rule today: Don't worry about saying anything you think Whitley or I may want you to say."

"Or I may want me to say."

"Don't worry about it. Don't try to decide if it fits or doesn't fit."

"I'm not worried about that. What I'm worried about is more unconscious motivations. Consciously I'm not going to do that."

Budd Hopkins: "Try not to censor, not to judge, is what I mean."

Dr. Naiman: "Not only will we not hold you responsible for your unconscious motivations, we welcome them. Just let your unconscious run free. There's nothing wrong with that. We want those associations. You look kind of puzzled."

"You mean if unconsciously inside I say, 'Darn it, everybody else saw light, I really want to see light—'"

"That's not so unconscious! That's very conscious! That's what was operating last week."

"That's like if there was an auto accident and everybody else got the number of the car and you didn't. You feel like a jerk."

(She was then put into a trance. For a time she reflected on a dream of a big, beautiful Victorian house on a grassy hill. It soon became evident that this was no symbol for a flying disk but rather for our family life.)

"OK, are we ready to leave that dream now, of that white Victorian house?"

"Yes."

"OK, if it's agreeable to you while you're in this trance, I'm going to change places with Budd and he'll take over questioning."

"Yes."

(All questions that follow were asked by Budd Hopkins. There were further questions about the "little white thing," which elicited the opinion that she did not remember it from her childhood after all.

Hopkins, moving to the night when the white thing had appeared in our apartment, tried to get her to describe any feelings she might have had about it.)

"I thought it was very odd that it had revealed itself, because if it had just poked Whitley it would have been just one of those odd things that Whitley says, and I'd say, 'Well, it was just an experience he had.' But since it poked me and our son, it kind of gave him away, and I thought it was odd of him to do that. With poking all of us . . . he revealed himself. Even appearing on the fire escape to the sitter, we wouldn't have tied that in. We'd have thought she'd gone crazy. Or that it was a prowler, and we would have been worried. Even that wouldn't have done it. Even if our son had seen it too, I wouldn't have believed it. I would have thought he'd just been influenced. Except that he was poked. . . . If he hadn't been poked, if he'd just said he saw something like Casper the Friendly Ghost, which is what he said, well, you'd just think it was a dream. Even if child and father had the same dream—well, sometimes they have a kind of ESP together, and they always have had. So even that's interesting, but it's not something you can go much further with. It really interested me that it would give itself away like it did."

"Take a few minutes and think through a dialogue, an imaginary dialogue—questions you might ask, answers you might get. What Whitley or your son might say. . . ."

"I can't imagine talking to it. It doesn't seem like something that would talk. I mean, I can't—it just doesn't seem like something that would talk. They aren't capable of talking. I just can't imagine that. I just wouldn't even think of asking it anything. I don't get the feeling it wants to talk, and I don't get the feeling that it can talk, or that if it could it would necessarily want to communicate. I don't think it does."

"Just one last question about it, since we know he came there, and was seen several times. . . . Do you ever have any feelings that he was there any other time, any inkling, any sense of another time you felt he was there?"

"No. I know Whitley does because he sees things out of · the corner of his eye. That's why I think it made a mistake poking me. Because that gave it a kind of reality testing. Then appearing to the baby-sitter gave it further reality testing. It's almost like he was making a mistake there, wasn't thinking through his plans well enough."

"What do you think his plans were?"

"Well, that I don't know. Seemed impish to me, to poke people and run."

"So, we'll move on to something else. I want to go to the October fourth night again. And this is a strange night, and you've had glimmerings of things and half-memories. Describe hearing your son crying. I want you to take a few minutes and hear that sound, as if you were in bed that night, hearing the sounds, listen for the sounds . . . any words, what kind of voice . . ."

"I don't know if he said 'Mommy, Mommy' or 'Daddy, Daddy.' It seems like he screamed. It seems like he called me, but everybody says he called Whitley. Screaming, though. [Long pause. Becomes visibly tense. Gasps.] Well, I don't want to say it because I feel it's been influenced and I don't want to say it."

"Don't worry. Say what you feel."

"Well, I know Whitley told me it was him screaming. He told me that. Now, when I take that thought into my mind, and then I think about the screams, I can hear Whitley screaming. It's very hard, because Whitley's not the one who's supposed to scream. He's supposed to protect us. But I can hear him screaming. I can see his face, very frightened. Terrified. His eyes widen and get very white. Just so frightened. I don't know—is that real or not? Because maybe my imagination is doing that."

"Don't worry about that."

"If he was screaming it would be so unusual. He's always so calm. But he does get frightened. He gets very frightened sometimes."

"The way you describe his voice, his face—"

"Oh! I can picture it! I'm trying to remember when I would have seen it. I just hear the voice of a woman . . . he's so frightened . . . and I think at the same time he would have been a bit ashamed of himself, because whatever he saw, he would have been frightened for us, not just for himself. But he was so frightened that he had to feel mostly frightened for himself."

"Did he seem very far away from you when he screamed?"

"No, because I can see his face. No, not far away."

"Was he in the room?"

"That I don't know at all. I don't picture any room."

"Can you remember another time he screamed like that?"

"Well, I'm trying to picture if there ever was a time. There've been some times when he's seemed frightened, but I don't think there's ever been a time when he screamed. You know, it's frightening to see a man scream, because men don't scream. Maybe they should or could, but they don't. So it's an experience you don't have. You never feel a man scream. I think most men don't even know if they could scream."

"Why is he screaming?"

"[Whispers.] I don't know. [Long silence.] It's fading away now. I was trying to think about that time, what I remember about it."

"You said you heard a woman's voice? Annie?"

"Mumbling in a soothing way."

"Mumbling?"

"Yes."

"Do you remember words?"

"I don't remember them. Maybe it was the tone of voice that made them sound soothing. Saying like, 'That's OK, don't be afraid.'"

"Did it sound like Annie's voice?"

"It was deeper. She has kind of a highish voice. [Long pause.] I get the feeling of ignorance being a kind of protection for me."

"Tell us what you're feeling now, Anne."

"I feel like I don't want to say anything. I don't know why that is. Usually I say a lot of things. [Long pause.]"

"I want you to say what you feel. Can I ask you another question?"

"Yes. If you ask a question, I might actually talk."

"In all of this, can you tell how Anne relates to all this?"

"I know my role, and it's rather a tiresome role, but—born with a certain personality, you can't fight it. I'm the one who's not informed, except through Whitley. I'm the one who responds emotionally. I know if it feels right. Whitley doesn't have any talent for that at all. Sometimes he can't feel the most obvious things."

"Do you feel your roles have been chosen? Did you choose these roles?"

"I feel they're inevitable roles."

"Because of the person you are?"

"Yes. I also feel that they're roles not only because of who you are but who you're with, and therefore you play certain parts, according to who you're dealing with."

"I want you to take a few minutes and think over all of this, your role, your son's, Whitley's, the little white thing, Whitley's screaming . . . mull over these images and think, what is central, what is marginal, what does it mean?"

"A feeling that Whitley was vulnerable. That's a rather frightening feeling. I would rather not know about these things that make Whitley vulnerable."

"Anything else, Anne?"

"No."

She was then brought out of the trance.

"Whitley's supposed to go. They came for Whitley."

I listened to the recording of Anne's first hypnosis on March 17, 1986, the Monday after the "confirming" encounter in the country. I hadn't listened to it on the previous Friday because she told me she hadn't remembered anything much. And indeed, on careful questioning, that was her perception.

I asked her, "What do you mean, 'Whitley's supposed to go?'"

"Well, that's what I said."

"Do you see me go?"

"No. But I hear it. There's a lot of noise sometimes. I keep my eyes closed."

"But don't you worry?"

"No. You're always there in the morning."

Fortunately, by the time I did listen to the tape I had become so used to being shocked that I did not really react too badly. I didn't end up stalking the streets or sitting in my office staring into space.

But her testimony had a powerful effect on me. It was by no means a "typical abduction scenario" that could have been drawn from subconscious memories of things she had read in the paper over the years. It was unlike other testimony—and thus was almost certainly taken not from her cultural background but from her actual memories and perceptions.

Hers was probably the most remarkable element yet to be introduced into this account. This was because there seemed to be so much unconscious process implied by her testimony. It really did appear that she had performed a function she had been trained to do. And then there was that enigmatic female presence. In my own hypnosis I remembered it making some sort of noises to me when it was beside the bed on the night of October 4. Anne remembered this too, Despite the slip about my screaming, she had no reason to identify that presence at the bedside, or to add that it was saying something while the screaming took place.

The temptation was, of course, to say that the visitor hypothesis was now so compelling it must be true. Testimony like hers, supportive in a totally unique manner, suggested very powerfully that there was some sort of design behind our experience. They had been taking me for reasons of their own and Anne had somehow been programmed to rehabilitate me by regrounding me in life.

However, it seemed to me that a rigorously objective approach still might prove more productive than surrender to a specific view.

But how to remain objective? I was being *exposed* to this. I was disappearing into the night. I had remembered probes going into my brain. My wife had painted a picture of me as a sort of soldier of the night, vulnerable and helpless.

One could state a few things with certainty, if one was careful. Something happened to me and possibly to my son. Its source and nature were unknown, but there was a strong suggestion that it included some sort of physical component external to and independent of us. This could be anything from some sort of sensitivity as yet unknown to fluctuations in the earth's magnetic field to actual visitors. Another thing that could be stated was that my wife had been aware that something was happening, and she responded by preserving her own neutrality—maybe she had been trained to do this and maybe not. It could also be that she was doing it out of an instinct to help her husband. The support she had provided may have been her own invention, rather than the outcome of training or suggestion from the visitors. Could she herself have been the woman—or the source of the female being—who at once gave me those insights on the night of October 4 and comforted me in my anguish?

Who were the old gods, really? Perhaps we gave them to

ourselves. When unconscious was joined to unconscious, maybe this was one possible outcome.

In general, Anne's memories were clear until it came to anything that might have related to the visitors. At that point she became unable to remember. This was most forcefully illustrated early in the transcript when she was recalling her day alone with our son on July 30.

We have questioned him very gently about this matter, and have discovered a wealth of information, which I will deal with in a separate section. Before Anne's hypnosis I found two short essays he had written for his school journal over the course of the fall, both of which could easily be descriptions of events relating to the visitors—or they could simply be the work of an imaginative little boy. And yet even the drawings of the "monsters" accompanying the stories suggest the large, slanted eyes of the visitors.

As both stories concern only him and his mother, we decided that they might refer to July 30. Since the three of us are almost never separated, it was easy to pinpoint that particular date. I had gone to Philadelphia to appear on a National Public Radio program. I spent the night at the Harley in New York and returned to the country on the morning of the thirty-first. I found everything totally normal, and my wife and son perfectly happy.

Were it not for our son's two essays and all these other strange occurrences, we never would have even guessed that something might have happened that day. Before her hypnosis, nobody told Anne that she would be asked about it, nor was any allusion made about why. She was unaware of the essays in the journal, which we had prevented her from seeing.

She remembered her day clearly until she reached the evening. Then she seemed to think that the two of them might have been invited somewhere. Then she went almost totally blank. Both of our son's essays refer to her fainting when the monster appeared.

Interestingly, she remembered watching "TV" at some point. I remember more than once watching a screen, such as the gray one I was put in front of when I was twelve.

Hypnosis then proceeded to a regression about the night of October 4. Neither hypnotist nor subject knew much about the events of that night, as is clear from their initial mutual confusion.

Frankly, Anne's totally unprompted allusions to a vague and powerful and very definitely female presence have been one of the things that has left me with long thoughts. I have gone to her and watched her in peaceful sleep, and wondered what it all might mean.

When she was first asked by Dr. Naiman what she remembered about the night of the fourth, she evidenced obvious distress, screwing up her face and clenching her eyes as if shrinking from a painful sight or noise. And yet when he asked her what she was thinking of, she promptly replied that she didn't know. A little persistence on his part brought a strangely conflicted memory of a night of activity that went on around her but in which she was not allowed to participate. At first she clearly remembered that the night was uncomfortably light, although she later denied this memory. As Dr. Naiman had not been apprised of the importance of the light, he made no special effort to draw information about it out of her, thus leaving both her memories and her denial intact. This also means that there were probably no hidden cues that she should recall the light more clearly.

After the session, she was asked what had made her say that the night seemed too light. "I had a vague memory of my eyes closed and my eyelids all lit up as if the light was on in the room. But it was very vague."

She was asked why she repeated so many variants on the theme that it wasn't a peaceful night. Despite reinforcement during hypnosis that she would shake some of these memories free afterward, she was not able to do so. She said, "I feel like I'm a piece of spaghetti with you on one end pulling and them on the other end refusing to let go."

She finally closed this section of her regression with, "I just see a light. I mean, it's not dark. You know, it's not dark."

Later in the regression she began to make references to

the house being full, as if there were something "different" about it, to use her word. We often have houseguests in the country, and the presence of Jacques and Annie was nothing unusual. Was she trying to indicate that somebody else was present in the house? The transcript was not suggestive enough on this point to be sure, but during both sessions she indicated that a woman was present. There was also that cryptic reference to "a friend" being in our bedroom, a reference that was never expanded upon.

When asked who this friend might be, she said she just had the feeling that somebody was there. Why *friend*, though, why not simply *person*?

"It was somebody we knew. An old friend."

"Jacques or Annie?"

"No. Somebody else."

"Can you picture them?"

"No. It's just what I felt."

Then there was the matter of who screamed. We carried out experiments at the house to find out just how clearly a voice from our son's room could be heard in our bedroom above. Screams could be heard easily. But loud talking was much less audible, and it would not have been possible that soft words of comfort could have been heard over screaming, even given our sparse soundproofing.

However, if the screaming was actually much closer to Anne, the words of comfort would also have been audible—especially if they were intended for us both and the screaming was muffled by some unknown effect.

There followed the first of the allusions that supported the notion that I ought to revise my understanding of my life. "I don't think Whitley was there very much. He was gone. You know, he goes sometimes at night. He goes and works. Or he just goes."

I don't *remember* going, though. I never work in the middle of the night. Once I'm in bed, I generally stay there all night unless I hear our son. And that happens no more than two or three times in a year.

While Anne's hidden role seemed to be that of passive

supporter, her own life role is very different. It was clearly revealed by a statement she made before the second session, when Dr. Naiman asked her if her presence in his office was voluntary. "If I had said, 'No, I'm not going through with this,' I wouldn't be here." She is as independent a person as I know, a committed feminist who is politically and socially as active as she cares to be. Except when it comes to this. In this matter, she is passive, which is in itself awfully strange.

As the intensity of the experience built, Anne became uneasy with her role. "Things were going on and I wanted to know what was going on!" Her tone became forceful, almost angry.

When asked why she didn't simply go and see, she repeated that she wasn't supposed to. Supporting this came the first of a number of what she feels are references to a female authority: "It was like your mother said to you, 'You have to stay here.'"

Anne's hypnosis strongly suggested that I'm taken all the time. And mine as well implied more than the two recent occasions. When she was being hypnotized Anne had no idea at all that I remembered more than two occasions when something strange happened. So why did she say "friend," as if a familiar individual were present, and why did she assert that I go "all the time"?

When Dr. Naiman and Budd Hopkins moved to the events of December 26, there was a flavor of what it must be like living with all these strange secrets when she made reference to my talking about the crystal in the sky. I remember the image clearly, and I remember being nonplussed when I spoke of it, because even at the time it seemed like a sort of falsehood—something I needed to say in order to put some deep uneasiness to rest.

She said frankly that she did not consider me a "down-to-earth guy." I'm glad of that; after all that appears to have been happening, she would have to be incredibly imperceptive to think that I was down to earth. Dr. Naiman, quite naturally, asked her if she thought I should go to a psychiatrist. Her reply was interesting: "No. Because he—I think he can deal with these problems."

What? I'm seeing things, claiming to fly around rooms, and my practical, no-nonsense wife *doesn't* think I should see a psychiatrist? Perhaps she knew that there would be no point, because on the level she would not directly address, she was aware that these are the side effects of real experience.

I will recount briefly the incident of "flying around the room." In March or April 1985 I was lying in bed in the country house, reading a book, when I suddenly had the feeling that somebody was in the room. I was confused, because the room seemed empty. It seemed almost as if there were somebody there who was able to remain just at the corner of my eye. The next thing I knew, I floated right out of the bed. I did not tell Anne that I saw a swirling, dizzying jumble of trees, house, and moon right after that. It just seemed too odd, so I contented myself with saying that I had seemed to float around the room. Flying dreams are not unusual, but dreams that vivid that take place when you are reading, not apparently asleep, are awfully hard to accept, which was why I mentioned it to her. I needed to talk about it. And there she was, ready to play her assigned role. Instead of asking if I thought I'd like to see a doctor, she just laughed and continued to act as if everything were totally normal, which was enormously reassuring, and I soon forgot the incident.

Anne's regression became a little confused at this point, because Budd Hopkins made the suggestion, "Back to that night," without specifying which night.

She thinks she then confused the nights. "It was like a party. There are lots of things going on here now." When— October 4 or December 26? She does not remember, although she states that Jacques and Annie weren't invited, so that may mean the twenty-sixth, when they weren't there.

Yet again there was reference to the mysterious female authority figure: "It's like your mother says, 'No, you can't go.'"

Finally she volunteered that she's often felt that there are things "going on" with me that she wasn't "supposed" to know. She then revealed a definite role: "I'm supposed to kind of help him afterwards to deal with it. That's my role. But I can't stop them, you know. He just has to."

She was then specifically asked if I have hallucinations. Her reply was that I do not have hallucinations, but "they come to him because of his head."

She then related her perceptions of the "little white thing" that invaded our apartment in the Village. What it was we will probably never know, and I cannot even guess its purpose.

On listening to the tape of her hypnosis, Anne felt that something seemed to be missing, and found it odd that she had remembered so little about the crucial periods of time— or so she thought. It appears, on careful analysis, that she remembered a great deal.

There was another reference to "the voice of a woman." She also admits that it was not Annie Gottlieb's voice, although not by saying so directly. "It was deeper. [Annie] has kind of a highish voice."

There is another possible explanation for Anne's testimony. It could be an expression of faith for a man she deeply loves and desires to protect even from the toils of madness by a subtle act of confirmation, really a hidden communion, an indirect sharing—of an experience she did not have enough information about to confirm in convincing detail.

One night in April she talked in her sleep. I had thought to call this book *Body Terror* because of the extreme physical sensation of fear I had felt on December 26. Suddenly she said in a strange basso profundo voice: "The book must not frighten people. You should call it *Communion*, because that's what it's about." I looked over at her intending to say why I thought my title was better, and saw that she was totally asleep. Then I realized where I have heard that voice before.

I went to my wife and looked down at her sleeping form, my mind full of question and wonder.

Our Son

We have been careful to preserve our child from the faintest suggestion that he might be dealing with something outside of normal experience. We have told him that he has had some scary dreams. Oddly enough, he seems to take this notion to be a sort of adult fantasy. His own descriptions of what he remembers are completely straightforward, and he doesn't characterize them as scary.

While he is more than willing to call them dreams if we want him to, he seems equally comfortable with the idea that they are memories. This exactly parallels my own perception: The material has the taste of real memory, and yet it is so strange that it also seems like a dream.

I have asked my son to describe any strange dreams he recalls. He has never been hypnotized and he won't be until he can decide for himself if he wishes to do it. No matter what the source, this material can be very disturbing indeed under hypnosis and it is certainly not the business of a parent to assault a child's mind by such experimentation.

Here are some of my son's dreams, in his own words.

"Well, I was dreaming that I was on a boat with Ezra [a friend of his] and someone was attacking and we were about to dive off and I was in the middle of the air when I switched to this dream where I was in the hospital in the future where they were trying to cure some kind of disease. I'm not sure what it was. And I was taken out of my bed and onto a cot and out on the porch."

"Who took you out of your bed and onto the cot?"

"Some kind of doctor."

"What did he look like?"

"Oh, he was a very short and fat man with glasses that came out pointed upward like that. [Gestures as if eyes have a pronounced slant.] And he always has a big fake smile on him. [Smiles from ear to ear with his mouth closed.] He kind of kept it there except when he was asleep."

"How did you know he was asleep?"

"Well, he had—well, that's because he worked in the night and slept in the day."

"What did his eyes look like?"

"He was wearing regular glasses. His eyes were a kind of greenish-blue color. Dark. The only two faces he had was this. [Again demonstrates smile.] And then a small one when he was sleeping. [Makes an O.]"

"Mouth open?"

"Yeah."

"When his mouth was opened, it was round?"

"Yeah. Puckered. Big puckered."

"Did you see him when he wasn't smiling?"

"Yeah, when he was doing the operation on me."

"What kind of operation?"

"Well, it was kind of like a test."

"What did he do?"

"It was a disease on my arm."

"He did something to your arm?"

"No, wait. He kept your nose cold like when you eat a lot of ice cream."

"Did it hurt?"

"No, not really."

"You say you were examined on the porch. What do you mean by that?"

"Well, they took me onto the porch. There was no way to get me into an operation room because of all the moving equipment. And then by the porch light I mean kind of like the outside lights at the country house. You know at the country house there's that porch light?"

"Yeah."

"That's the light that was on. Then they took special lights and examined my nose and took X rays and stuff." (This last statement could easily be a buried memory of a babyhood injury to his nose, which involved an X ray to determine whether it was broken. But this memory seems to be mixed in with other material of a totally different nature.)

"What kind of lights?"

"Some were blue lights and they would look through the front of it and the blue light would make them see through me without an X ray."

"Yes?"

"And there was an orange light that was supposed to see not my bones but the inside skin and what was happening. Instead of having X rays and stuff, they had lights. They had big lights. Green lights."

"Ever remember a dream where a monster came in the house and Mommy fainted? What's that from?"

"That's one of my journal stories."

"Yeah. Why'd you make that one up for your journal? Do you know?"

"I don't know. Not really. I remember it vaguely. Because I wrote that one a long time ago." (Early fall. It was now March.) "It was free journal story period and I couldn't think of anything. I was tossing and turning in my desk trying to think of something. And then suddenly that dream just popped in my head."

"What was it like, that dream?"

"I was in a—I didn't explain it totally on the journal. It was in a cornfield, my mommy and me, and I was chewing on a piece of corn and my mom was telling fairy tales. And then suddenly this big, big—about let's say from the lobby of this building up to the top—hovered over us. It was colored orange, green, had blue feet." (Orange and green are colors associated with lights on the flying disk that has been seen in our area.)

"It was a thing, like an animal or a creature?"

"It wasn't like anything. It was just this big, massive thing. It had these big bumps all over it that were blue and its feet were orange—"

"Do you suppose you were seeing something flying over you that was blue and orange and green, and you were confused as to what it was?"

"It was like it was flying. Kind of."

At this point I felt that I had made a mistake with my last question, in that it was so heavily weighted with suggestion. I concluded our conversation by reassuring my son that he'd had some really neat dreams that were very interesting to hear about.

He then went about his afternoon business, reading *Tin-Tin* and making a St. Patrick's Day card for his grandmother.

I sat in my chair, haunted by what my son had said. Most particularly, I thought of the incident in the cornfield. I will relate a dream I had had shortly before we spoke.

The three of us were together in the English countryside in my dream. We had rented a cottage. The inside of the cottage was identical to our cabin. I was confused, because Anne and our son were not there and it was already the evening. I was sitting up in bed when I got a call on the phone. I remember saying to the caller, "No, it's all right, they're just staying out all night." On some level I was full of fear, but on another I seem to have accepted their disappearance by justifying it to myself.

In the middle of the night there was a knock at the front door. I opened it to find my son in the company of a group of "rescue workers," ordinary men and women with deep, soft, and loving faces. My son was naked except for a dark blue cap that one of them had put on his head. He was moving strangely, as if he had no control over his own muscles. His eyes looked as if he were in some sort of trance. I gathered him in my arms, because they told me that touch and hugging would bring him back to normal. Then I looked around for my wife. They shook their heads sadly, and the care and love radiating from their eyes was such that I was not bereaved but reassured that she would be back soon.

Then I was abruptly transported to another place. I was given to understand that Anne and our son had been found here, hiding. It was a cornfield, just like our son's dream.

At bedtime that night he wanted to talk more about dreams. I did not record our conversation, but he complained of two things. The first one was that when he started to go to sleep, his whole body would tingle and he would feel as if his hair were standing on end. A voice would then ask him about his day, how he felt, and "private things" which he did not wish to discuss with me.

He also complained that he saw a skeleton looking at him when he was trying to relax. The conversation went as follows:

"A skeleton?"

"Yes, and it keeps staring at me like it was right in front of my face and it won't go away."

"What does it look like?"

"Well, it's—oh. It's not a skeleton, it's one of the thin ones that stood around behind the doctors."

"What thin ones?"

"You know, the thin ones that are always saying 'We won't hurt you'? Them. It's not a skeleton, it's one of the thin ones."

The appearance of these people has never been discussed with my son at all, not by anybody, and yet his description of short ones and taller, thin ones is not only consistent with my own observations, it is consistent with the experiences of many of the other people who have encountered the visitors.

He had bought a book of haiku at the Strand used-book store that afternoon, a book entitled *A Net of Fireflies*. I did not tell him that I had bought the same edition when I was twenty and living with my grandmother, and derived immense pleasure and comfort from it. He wanted us to read haiku to one another. I read:

> *With tender impact on the icy air,*
> *The peach-buds burst: their silken petals flare.*

He smiled his huge smile and commented, "That was really a lot of pictures for so little words." Then he read:

> *Without a sound the white camellia fell*
> *To sound the darkness of the deep stone well.*

Afterward he said, "Dad, you know, we like the haiku and all the beautiful words. But the thin ones, it's like they *are* the haiku. Inside, they are haiku."

That night a father stayed a long time with his child, wondering about the soft fire of communion that might be hidden between the breaths of his life.

SIX

No building ever came into being as easily as did this temple—or rather, this temple came into being the way a temple should. Except that, to wreak a spite or to desecrate or destroy it completely, instruments obviously of a magnificent sharpness had been used to scratch on every stone—from what quarry had they come?—of an eternity outlasting the temple, the clumsy scribblings of senseless children's hands, or rather the entries of barbaric mountain dwellers.

—FRANZ KAFKA, "The Building of the Temple"

A STRUCTURE IN THE AIR

Science, History, and Secret Knowledge

What Is Going On?

In the past forty years the question of the real nature of the UFO experience has been addressed by psychologists and psychiatrists, most notably Carl Jung, by public personages ranging from presidents Jimmy Carter and Gerald Ford to Senator Barry Goldwater, by a plethora of scientists, by the United States Air Force and various security agencies, and by the general public.

In addition to all this recent interest there are reports of disk sightings, of airships, of little men in silver clothing, dating back over a thousand years.

I began this search by assuming that I was dealing either with a mental aberration or a visit "from another planet." If I had been asked, I would have said that the nature of my experience indicated that the visitors hadn't been here too long, and that I had been studied by a team of biologists and anthropologists.

Judging from the extraordinary range—not to mention the age—of some of the material I have found, this cannot be the whole answer. Even if partially true, there is much more to it than a recent arrival of more or less comprehensible visitors. Whatever this may be, a correct and final understanding of it certainly poses an intellectual, emotional, and spiritual challenge of unprecedented complexity and subtlety.

There is so little final knowledge of this phenomenon that it is impossible to do more than speculate about its actual nature. But speculation need not be random; it can be careful and directed.

The visitors could be:

- from another planet or planets.
- from earth, but so different from us that we have not hitherto understood that they were even real.

from another aspect of space-time, in effect another dimension.

- from this dimension in space but not in time. Some form of time travel may not be impossible, only unlikely and probably very energy-intensive. For example, if we could convert a human being into some sort of energetic medium—say light or radio waves—then place a reconverter 100,000 light years from earth, a person could step through a door here, feel as if he had come out the other side instantaneously, then step back and find that he was 200,000 years in the future. A cumbersome time machine, but it would work. We cannot assume that time travel is out of the question.

- from within us. I keep returning to this hypothesis because I find it so endlessly interesting and at its core so compelling. I suppose the idea that the gods we create would turn out to be real because we created them has a certain ironic appeal to a modern intellectual.

- a side effect of a natural phenomenon. We know so little about how magnetism and extra-low frequencies of all kinds affect the human organism. Perhaps there are natural electromagnetic anomalies that trip a certain hallucinatory wire in the mind, causing many different people to have experiences so similar as to seem to be the result of encounters with the same physical phenomena.

- an aspect of the human species. We have a very ancient tradition of afterlife. The respect with which Neanderthals buried their dead in the Middle East more than thirty thousand years ago suggests that this belief may actually predate our own species. Maybe we *do* have an afterlife, but not quite in the way tradition suggests. Maybe you and I are larvae, and the "visitors" are human beings in the mature form. Certainly, we are consuming our planet's resources with at least the avidity of caterpillars on a shrub.

Ancient astronomers of India believed that the Siddhas (human beings who have attained perfection) revolved between the clouds and the moon, having been transformed into a lighter, less material state.

Maybe the ancient and revered concept of human spiritual transformation relates to the emergence of the adult from the larva.

In our society *transformation* has a bad name, having been associated with various meditation fads and instant success groups. But real transformation has nothing to do with gaining a better life in this world; deliverance does not involve trying to use Buddhist chanting techniques to acquire a new Mercedes, nor is salvation a side effect of Fundamentalist healing services. Transformation for a Zen monk, a Moslem sufi, a Catholic, or a Jehova's Witness is the same: It is a matter of delivering one's self into the possession of God. Meister Eckhart puts it very well when he says, "We must become as clear glass through which God can shine." But this involves giving up the "self," which feels just like dying.

My historical survey has found that the core experience of seeing flying disks and small figures goes back a long, long time. If we are dealing with extraterrestrials, would they really have been here, say, for a couple of thousand years and remained hidden all that time? Or did they come recently, and find a way to slip themselves into a preexisting human mythology in order to hide there? Or even more extraordinary, it is possible to consider that they might have really, physically arrived sometime in the future, and then spread out across the whole of our history, in effect going back into time to study us. This might mean that they could be here only a short time—say a few weeks or months—but are carrying out a study that would seem to us from our position in sequential time to have extended over our entire recorded history. Of all the theories, this one alone explains why creatures in 1986 would seem so unaware of our languages, nature, and even our clothing, and yet possibly have a history as fairies and gods going back thousands of years. It might also explain why they are so enigmatic and seemingly immaterial. If time travel can exist and is involved, God only knows what the travelers would look like to us as they reached back from the future. From our perception, they might appear in the skies one moment and have full understanding of our entire history, our cultures, and our languages the next. This would be because they would travel back across the whole of our time in the few moments between discovery and understanding, picking up and assimilating all the details at their leisure, while we would be presented with the illusion that they already knew us well even though they had, in effect, just arrived.

All these hypotheses are interesting, but none of them is in the least provable. They do suggest, though, the rich range of possibilities that are available for further study.

The central question remains, Is there anything real about all this?

I have spent a great deal of time in the past few months searching what could only be described as a morass of literature on the subject of unidentified entities and their craft. I have talked to scientists who think that it's all nonsense and to scientists who are not so sure. I have read dozens of case histories and met many other people who have had visitor experiences.

Something is happening. This is clear. It is not a version of known phenomena.

In my research I found an undertone of claims that the government knew more about this matter than it was saying. I decided to do some investigation into the truth of this possibility.

I found myself in a minefield. Real documents that seemed to be false. False documents that seemed to be real. A plethora of "unnamed sources." And drifting through it all, the thin smoke of an incredible story.

There is some small reason to speculate that the United States government may have had some sort of communication from visitors as early as the late 1940s, as well as obtained pieces of crashed disks and the bodies of the occupants.

I base this notion on the two best pieces of evidence I

could find, both of which I have personally investigated and confirmed as genuine.

These two pieces of evidence are certainly real, in the sense that they are not witting hoaxes. Of course people can make mistakes. The first document is a letter written by Dr. Robert I. Sarbacher, dated November 29, 1983, addressed to Mr. William Steinman, a UFO researcher who had inquired about Sarbacher's government consulting activities during the late forties. The letter has been offered for publication, but to date has not been fully exposed except in the journal of the Mutual UFO Network, a group of people, many of them scientists and academics, interested in serious study of the phenomenon. The letter was also referred to and quoted in *Omni*. As Dr. Sarbacher died on July 26, 1986, a few days before I became aware of his letter, I was unable to interview him, but I discussed his case with Mr. Barry Greenwood, co-author of *Clear Intent*, who had had extensive discussion with him. These discussions revealed that Dr. Sarbacher did not appear to know more than he stated in his letter, but he was quite certain that the facts he did relate were as he remembered them.

Dr. Sarbacher was a Department of Defense Research and Development Board consultant during the Eisenhower administration. Educated at Johns Hopkins, Harvard, and Princeton, he is the author of *Hyper and Ultra-High Frequency Engineering, Research Accrediting at Military Establishments,* and the *Encyclopedia Dictionary of Electronics and Engineering*. This last book is considered a fundamental contribution to science.

He has been dean of the Graduate School of the Georgia Institute of Technology and has participated as a consultant for the Oak Ridge Institute for Nuclear Studies, in Tennessee, and the navy, as well as the Department of Defense. He has held many corporate directorships, among them as director of the General Sciences Corporation and the Union Life Insurance Company.

Dr. Sarbacher writes, in part:

227

Relating to my own experiences regarding recovered flying saucers, I had no association with any of the people involved in the recovery and have no knowledge regarding the dates of the recoveries. . . .

About the only thing I remember at this time is that certain materials reported to have come from flying saucer crashes were extremely light and very tough. I am sure our laboratories analyzed them very carefully.

There were reports that instruments or people operating these machines were also of very light weight, sufficient to withstand the tremendous deceleration and acceleration associated with their machinery. I remember in talking with some of the people at the office that I got the impression these "aliens" were constructed like certain insects we have observed on earth. . . .

I still do not know why the high order of classification has been given and why the denial of the existence of these devices.

All I can say is that I remember very well the eidetic image. I described the joints of the creature I saw as "insectlike." And my hypnotic transcripts continually refer to my impression that I was dealing with creatures that moved like insects. I did not know of Dr. Sarbacher and his letter until August 9, 1986, months after I had my experiences.

The combination of Dr. Sarbacher's recollections and my own memory of how the visitors acted and treated me may provide insight both into why they are so secretive and why they have been so seemingly indifferent to our rights and dignity with their forceful abductions.

If the visitors are indeed insectlike they could be organized as a hive, and not only as small as I saw but also as light as Dr. Sarbacher recalls being told. They may be no physical match for us, not even in fairly large groups. In addition, it could be that there is very little sense of self associated with individual members of their species. Taken together they might be very formidable, but separate individuals may be almost negligible. If their mind is also a hive structure, it could be that their language is more a biological function than something learned. Maybe it is like the language of earth's hive insects: a complex combination of movements, smells, and aural output. There may even be more to it; one of the greatest of biological mysteries is how hives function, and whether or not a hive can have a group mind.

I will briefly illustrate the profundity of this enigma. There has been a study of bees under way at Princeton for some years. A part of this study was designed to determine how quickly a hive could find a food source if it was moved. Each day the food source was moved a measured distance. It was soon discovered that the bees would be waiting for the food source *at its anticipated location* before it was moved.

What a truly intelligent hive mind might have achieved, and how it communicates with itself and others, may be very hard to know.

My second piece of evidence that the government may know more about this than it is saying is a small but telling one, a press release that was issued in July 1947. In that month an incident took place on a ranch near Roswell, New Mexico, that is the granddaddy of all the "government has a UFO/alien bodies" rumors. Throughout 1947 there were many reports in the New Mexico newspapers of odd lights. But the first modern flying disk sightings, near Mount Rainier, Washington, had been heavily publicized at the time, and some of the New Mexico sightings may have been a result of confused observations of two V-2 launches at White Sands. One took place on June 12 and another on July 3.

However, on the evening of July 2, something was seen by members of the public in the skies over Roswell and reported locally. This object was brightly lit and crossed the skies in a northwesterly direction. On July 8 the Roswell Army Air Base issued a release which was published, among other places, in the *San Francisco Chronicle*. The release was issued by public relations officer Lieutenant Walter Haut, on order from the base commander.

> The many rumors regarding the flying disc became a reality yesterday when the intelligence office of the 509th Bomb Group of the Eighth Air Force, Roswell Army Air Field, was fortunate enough to gain possession of a disc through the cooperation of one of the local ranchers and the sheriff's office of Chaves County.
> The flying object landed on a ranch near Roswell sometime last week. Not having phone facilities, the rancher stored the disc until such time as he was able to contact the sheriff's office, who in turn notified Major Jesse A. Marcel of the 509th Bomber Group Intelligence Office.
> Action was immediately taken and the disc was picked up at the rancher's home. It was inspected at the Roswell Army Air Field and subsequently loaned by Major Marcel to higher headquarters.

Later a report was issued to the effect that this disk was a crashed weather balloon. Maybe it was, although it is odd that a crashed balloon would have retained enough of its tension to appear to anybody, even briefly, to be anything other than a flaccid plastic bag and some tinfoil.

Under the headline DISC SOLUTION COLLAPSES the *Chronicle* reported that General Roger M. Ramey had stated that the wreckage was "a high-altitude weather observation device" that "consisted of a box kite and a balloon." Later in the same story General Ramey is reported to have said that it was a "star-shaped tinfoil target designed to reflect radar." Since General Ramey claimed to have the debris in his office, it is strange that he could not seem to settle on an identification for it. Weather balloons, box kites and radar stars were all familiar to air-base personnel in the late forties, and Air Force officers on active duty would have been unlikely to mistake them. Even less likely to be confused on this matter would have been a working base commander such as Colonel William Blanchard, whose order led to the base press officer issuing the original press release claiming that the debris was a crashed disk. The debris was picked up by air intelligence officers Jesse Marcel and "Cav" Cavett—two more men who would have been unlikely to misidentify items like weather balloons and radar targets.

It seems possible that the debris may have been what the officers who originally saw it thought it was—and what they naturally told the press. They were obviously ignorant of General Ramey's cover-up plans.

Much has been made of the fact that the original discoverer of the device, rancher W. W. Brazel, was quoted by the Associated Press as being "sorry he told about it" and describing what he found as tinfoil and other debris. There are three enormous problems with the credibility of the rancher's statements. The first is that he was held incommunicado for days before he made them. The second is that members of his family have asserted that he made them under duress, and his being held incommunicado, which is a matter of record, would certainly support that assertion. The third problem is more telling because it does not rely on the rancher's verisimilitude at all. It is that none of the officers originally involved ever thought that they were dealing with the remains of known objects or they never would have allowed their press release to be issued in the first place.

On looking back through old copies of the *Skeptical Inquirer,* I discovered an article in the April 1986 issue entitled "Crash of the Crashed Saucer Claim." As it refers to the same incident I have been discussing, I researched it in some detail. Its primary assertion, I am sorry to say, was not supportable at all. The article claimed that the whole business began with a story in a book by Frank Scully entitled *Behind the Flying Saucers* (New York: Henry Holt & Co., 1950). I obtained a copy of this book and discovered that there is no relationship whatsoever between its rather dubious tales and the affair in Roswell in 1947. And I also obtained a photostat of the *San Francisco Chronicle* article from which I have quoted. The photostat is dated July 9, 1947, a year before the "events" covered in the book. To connect the book and the earlier incident in any way strikes me as an example of poor scholarship and seems not to be supportable. In addition, the *Skeptical Inquirer* article makes much of rancher Brazel's statements, as quoted by the Associated Press, but does not acknowledge in any way that he made them after a period of being held against his will, and in the presence of the Air Force officers who had been interrogating him. If all he found was a crashed weather balloon, why did they interrogate him? A more logical course would have been to interrogate and severely discipline the officers responsible for the "false" press release. There was no reason to hold an innocent citizen who had made an innocent mistake. Strangely, based on the service information I was able to obtain from the Air Force, I found no evidence of disciplinary action taken against any of the officers involved in the press release.

Was Mr. Brazel interrogated in order to induce him to change his story? There seems to be no other logical conclusion. The inescapable fact of the Roswell affair is that a group of professionally competent Air Force officers caused the publication of a press release claiming that the Air Force had recovered a crashed flying disk, after observing the debris. Only *after* this release was published was any attempt made to change the story. Even then, neither the release nor the professional competence of its author, his base commander, or the concerned intelligence officers was ever called into question, publicly or—as far as I was able to discern—in internal Air Force procedures. Instead the original witness, who was in no position to know what he had actually seen, was placed under duress and compelled to change his story. Even this process took a substantial period of time, which suggests that the man may have been clinging to the truth despite the frightening situation he was in. The recent attempt to debunk this story in the *Skeptical Inquirer* was not satisfactory. As a matter of fact, its author, it turns out, takes the unusual position that *all* unidentified-flying-object sightings can be explained. I have not found many scientists willing to make such a strong assertion about these transitory and poorly understood phenomena, and I wonder if the *Inquirer* has not stepped beyond the limits of healthy skepticism in its recent article.

That there are deep secrets connected with the area of unidentified flying objects cannot really be denied. In the 1970s Senator Barry Goldwater was denied access to secret documents concerning apparent research into UFOs being conducted at Wright Patterson Air Force Base. The National Security Agency has gone all the way to the Supreme Court to protect some of its documents about the disks.

The book *Clear Intent* by Lawrence Fawcett and Barry J. Greenwood contains legitimate documents obtained under the Freedom of Information Act that make it essentially impossible to contend that government personnel have not, at the least, had some very strange experiences over the years. It is essential reading because of its coherence and its clarity. Fawcett and Greenwood prove that some extraordinarily strange things have happened, and that the government has kept these things secret.

In 1966 a "Scientific Study of Unidentified Flying Objects" (the Condon Report) was undertaken at the University of Colorado. When it was issued in 1969, the Condon Report was instrumental in causing me to lose what small interest I had in flying disks and related phenomena. I read the preface, and saw that the project leader apparently thought that there was nothing of interest going on. From that I concluded that flying saucers weren't real and forgot about them.

Recently, I read the Condon Report more carefully and discovered that the internal conclusions are at variance with Condon's preface! His putting his thoughts at the front had the effect of hiding the actual realities of the report. It clearly states that a significant percentage of flying disk cases remain unexplained.

At the inception of the Condon Report, Robert Low, the

business administrator of the University of Colorado at the time, wrote to his superiors the following memo:

> Our study would be conducted almost exclusively by non-believers who, although they couldn't possibly prove a negative result, could and probably would add an impressive body of evidence that there is no reality to the observations. The trick would be, I think, to describe the project so that, to the public, it would appear a totally objective study, but to the scientific community, would present a group of non-believers trying their best to be objective but having an almost zero expectation of finding a saucer.

At the outset of the project, Condon told a public meeting: "It is my inclination right now to recommend that the government get out of this business. My attitude right now is that there is nothing to it. But I'm not supposed to reach a conclusion for another year." That is an unsound position to take at the inception of a study. Upon saying that, he should have resigned as project director.

Many of the scientists who participated in the study disagreed with Condon, especially after they saw the data. Some resigned in protest. One of them, Dr. David Saunders, published a book about it called *UFOs? Yes!* a few weeks before Condon announced his negative conclusion in November 1968.

The Condon Report ended the public interest of the United States government in the whole subject of unknown flying objects. Subsequent to the government's turning away from study in this area, very quietly the number of cases of people being taken by the visitors seems to have begun to rise.

Scientists nationwide have responded to the government's public position by refusing to take the matter seriously. Many people of the highest reputation have been sucked into this stance.

When the policy of denial was instituted, I doubt if anybody ever dreamed that the visitors would one day start marching into the homes of America in the middle of the night. But it appears that this may be happening. If so, then the public has ended up on the front line. And the visitors are not only entering our homes, they are entering our brains. And we do not know what they are doing to us.

It is not possible for there to be a more provocative or intimate intrusion. If, as seems clear, we cannot control the visitors in any way, then we human beings must create among ourselves a community of support. The only lasting damage I can find has not to do with any direct side effects of the visitors' activities, but with people being isolated with their experiences because of the indifference or incomprehension of competent scientific professionals.

The fact that respected public institutions such as the government and the scientific and medical establishments do not consider this a real problem hurts people, and hurts them badly. The lack of social support irrevocably isolates them when they need help the most. When they read false debunking stories or see others like themselves made the butts of jokes in the press, they are in effect assaulted a second time by their own society.

Cornell University professor Dr. Carl Sagan has stated many times that there is no evidence that unidentified aerial objects—and presumably visitors—exist. To be precise, there is no publicly acknowledged physical artifact. The large body of encounter memories, some heavily freighted with imagination, others more sparse, amount to an artifact of *something*. And there is a substantial body of carefully authenticated photographic evidence of the devices themselves that is very hard to refute in any way except on an emotional level. Of course, there are also liars who claim contact, and faked photographs—some of them skillfully faked.

It appears that there is more than a shred of evidence that there are visitors here, and that they are doing something that involves us. It is also obvious from their secrecy that they want very much to hide. Can it be that the government is inadvertently helping them do this, or even that they have somehow compelled it to act as it does? Certainly the combination of visitor and government secrecy has led to profound public confusion. We do not know what is going on. There is no publicly available reason to conclude that our earth is the object of visitation, or to support any of the other hypotheses that have been advanced. Indeed, any such assertions would be premature.

Maybe the visitor experience is what happens when the human mind looks into the mirror . . . and discovers that its own reflection is not only real but fearful to see.

Something is here. But what? And from where?

We come at last to the essence of the mystery.

Ancient Future

The first instance of an official attempt to explain flying disks, oddly enough, is not American. In Japan, General Yoritsune observed while on maneuvers that there were mysterious lights swinging and circling in the southern sky. The visitation continued through the night. In the morning the general ordered a group of scientific investigators to determine what had caused the strange disturbance. After consultation they announced that "it was only the wind making the stars sway." They are to be forgiven the profundity of their confusion, for the date of this occurrence was September 24, 1235, 751 years ago.

More recently, the distinguished Harvard astronomer Dr. Donald Menzel, in his 1953 book *Flying Saucers,* explained that a major sighting, carried out by professional observers with good equipment, was "an atmospheric lensing effect." According to navy physicist Dr. Bruce Maccabee, critical errors were made by Dr. Menzel in comparing his own lensing theory with the data reported by the observers. Specifically, the angle at which the observation took place was too great to allow lensing, even under Dr. Menzel's hypothesis. However, in his discussion of the sighting, Dr. Menzel did not mention the angle actually reported but assumed it was one at which it might have been possible to observe lensing.

The observation I refer to occurred at 10:30 A.M. on April 24, 1949, and was made by Mr. Charles B. Moore and a group of U.S. Navy trainees observing a weather balloon with a theodolite. Mr. Moore observed both an unidentified object and the balloon at the same time. He noted the azimuths and elevations of the object as it moved, and it was noted that the azimuths changed by about 190 degrees during the sixty-second sighting, and that the central angle between the initial and final sightings was 120 degrees. This information was filed with the Navy Special Devices Center.

I found that Dr. Menzel's description of the sighting in *Flying Saucers* is not the same as in the report. He claimed that what the observers saw was a mirage of the balloon, appearing at first above the balloon and moving straight downward until it was below and to one side of the balloon. But the report clearly states that the object appeared at first so near the balloon that Mr. Moore initially thought it was the balloon. The balloon remained in place while the object moved off to the north.

Dr. Menzel was well aware that a mirage cannot appear at a large angle away from the object that is the source of the mirage. In the appendix to *Flying Saucers*, Dr. Menzel calculated that the largest angle between the balloon and its mirage would be no greater than one fourth of a degree. But Moore's measurements were far different from that: To make the mirage theory stick, Moore's measurements would have had to have been off by about a hundred degrees. Nowhere in his book did Dr. Menzel mention the actual sighting angles reported by Mr. Moore.

Seven hundred fifty years, and it's still "the wind making the stars sway." That is much of the explanation that has been offered for flying disks and related phenomena by those who are emotionally or intellectually unable to admit the reality of the mystery. It is not because they are bad scientists, not at all. To a degree some of them may be shaking hands with elements of the intelligence community that have hidden information about the phenomena, but this is mere conjecture at this point. A far more compelling reason for this irrational behavior is suggested by a paper presented to the American Association for the Advancement of Science in 1969 by Dr. Robert L. Hall, professor of sociology at the University of Illinois: "We might describe the body of scientific knowledge accepted at any given time and the people who bear that knowledge as constituting an unusually strong belief system which resists inconsistent items of knowledge even more powerfully than a layman defending his political beliefs. . . . The very strength of our resistance to the evidence on UFOs suggests to me that there is clearly a phenomenon of surpassing importance here."

Since that paper was delivered there has been added a new element, which is that of scientifically educated people with Fundamentalist Christian religious beliefs. These "scientists" have joined forces with the debunkers, even founding official-sounding "skeptics' groups" that have Creationist motives.

The Institute for Creation Research has stated, "To date there is not one iota of real evidence in either science or the Bible that intelligent beings were either evolved or created anywhere in the universe except on earth. In any case, it is the planet earth which is the focal point of God's interest in the universe. There is no need to look, because there couldn't be anyone out there." The banality of this position makes it more pitiful than frightening, but there are competent scientists, such as Drs. John D. Barrow and Frank J. Tipler, who recently published a brilliant book, *The Anthropic Cosmological Principle,* which elegantly states a similarly unsound case, albeit from a much more intellectually substantial viewpoint than that of Creationism. The weakness of even the most sound "man-centered" case is our striking lack of samples from which to extrapolate predictions. We have one sample, and one only: this planet. If we could observe conditions on, say, a few million planets, we might be able to make more viable predictions, as we would then have a sample base comprising at least a small proportion of the probable planetary matter in the universe.

As the emotional charge of the debunkers and Creationists diminishes the impact of their position, so the paucity of samples reduces the vitality of more coherent man-centered arguments.

The truth is that we do not and cannot know the actual condition of life elsewhere in the universe because we are presently too ignorant of conditions outside our own immediate solar neighborhood. However, judging from the amount of evidence available, it may be possible to expand our knowledge simply by taking the flying disk and abductee phenomena seriously. We may or may not find visitors, but we would certainly find a body of data so compelling and multidimensional in its complexity that merely stating useful hypotheses about it is going to be a major challenge not only to the physical and behavioral sciences but also to the science and art of language.

This matter is a garden of luminous weed through which only a fool would dash yelling any doctrine at all, whether it be that of the Creationist and debunker or that of the UFO true believer. Even to approach the idea of the visitors, it is necessary to study a whole history of tall stories, bizarre tales, and—just possibly—truths.

It is our American habit to assume that there is something irrelevant—even a little silly—about the past. Our relationship to former times is expressed as nostalgia, not history.

When our government first started studying "flying saucers" in the late forties, it never even occurred to anybody official to consider having a look at the past.

Here are two stories:

In the little town of Merkel, Texas, on April 26, 1897, a group of people going home from church at night allegedly saw a heavy object dragging along the ground. They followed it until it bounced across a railroad track and caught on one of the rails. It was an anchor, tied to a rope. When they looked up, they saw an "airship" with lighted windows and a headlight on the front brighter than the light of a locomotive. Ten minutes passed, and soon a man was seen coming down the rope. He was small, and wearing a blue sailor suit. When he saw the people he cut the rope and the ship sailed off into the night, leaving the anchor behind.

The small beings I first saw were dressed in dark blue coveralls. This is not a unique description of the visitors' garb; perhaps it is a sort of night uniform. But then there are the kobolds, dwarfs who stalked the mines of medieval Germany and gave their name to the mineral cobalt . . . and cobalt blue. Why? They wore dark blue coveralls, too.

One Sunday in the borough of Cloera in Ireland the parishioners of the Church of St. Kinarius heard a noise on the roof. They went outside and saw an anchor embedded in the eaves. The anchor line rose up into the sky where there floated a ship on the air. A man leaped overboard and "swam" down to the anchor. After an altercation with the parishioners, he cut the rope and managed to return to the ship, which sailed away. The anchor remained in the church, but has since been lost, since this incident took place not in 1897, but around A.D. 1211.

What do these stories mean?

When people in the present time find themselves face to face with the visitors, they often think that they are among the first to see them. They do not remember what happened to St. Anthony of Alexandria, the founder of the Monastic movement, in A.D. 300. He was walking through an isolated canyon when he came upon a small figure, "a manikin with hooded snout, horned forehead and extremities like goat's feet." There was a brief exchange of words between this small creature and the saint, who ended the conversation by pounding his cane on the ground and announcing, "Woe to thee Alexandria, who instead of God worshippest monsters! Woe to thee, harlot city, into which have flowed together the demons of the whole world!" Needless to say, the creature fled.

During the reign of Pepin in the early Middle Ages, the

French were bothered by apparitions that were seen marching through the sky, camped out in tents on the reaches of heaven, and sometimes in "wonderfully constructed aerial ships" that flew past in veritable squadrons. People were annoyed at the presence of all this unquenchable grandeur and happiness, and both Charlemagne and his successor, Louis the Debonair, imposed penalties on the "Tyrants of the Air." As counterpropaganda, the sylphs kidnapped people and took them to their airy abode, showing them their world. But when the people were sent home, they were all burned at the stake without a second's hesitation. Presumably they had just enough time to scream out their stories before being engulfed in the flames.

It is not surprising to me that Marius Dewilde, a resident of Quaroble, a French village near Belgium, was so secretive after he encountered two dark alien figures standing near some railroad tracks in the middle of the night in September 1954. Mr. Dewilde, according to Jacques Vallee in his book *Passport to Magonia* (Chicago: Regnery, 1969), gave the French Air Police some calcinated rock from the site of the event, and it was handed over to an agency so secret that its name could not be mentioned by the Ministry of Defense.

Perhaps the keepers of these secrets across the world ought to reflect on the ageless nature of this experience. I wonder if St. Anthony's manikin and the medieval French Air Tyrants should join the calcinated rock in deep classification.

Perhaps this would be superfluous, as one "abductee" gave the calcinated earth from her backyard to Budd Hopkins, who has no access to classification procedures. Laboratory analysis of the dirt indicated that it had been subjected to intense heat. Dirt from the same garden had to be burned for eight hours at 800 degrees to achieve a similar effect, and there was no site evidence at all of a lightning strike or even of storm activity on the night the calcination occurred.

As the ages roll along, it could be that what changes is

not our visitors but our way of installing them in the culture. Maybe they did not come here in 1946, 1897, 1235, or even A.D. 300. I have reported that the being I have become familiar with looks like Ishtar. Maybe she is: She said she was old.

My point is that there may be far more to this than science or government or even religion can separately address. It would seem that our civilization is not paying attention to what may be the central archetypal and mythological experience of the age. If so, then this is the first time that man has simply refused to respond to the ghosts and the gods. Is that why they have become so physical, so real, dragging people out of bed like rapists in the night—because they *must* have our notice in order to somehow be confirmed in their own truth?

This may be primarily a matter of visions and chimeras battering at the door of physical reality. They are not simply flickering effects of the mind. Something is out there, and it wants in.

There are many instances of the surprising and subtle relationship between the visitor phenomenon and the hidden life of the mind. Understand, I am not presenting a hypothesis that denies that the visitors may be real beings from another planet and/or reality. All I am suggesting is that we do no know what they are, only that they are—and our relationship with them is very strange indeed.

During the great northeastern power blackout of 1965, actor Stuart Whitman saw an object and heard a strange whistling sound outside his twelfth-story window. He then heard a message to the effect that the disaster was "a warning to the world."

The first instance of an unidentified flying object causing a blackout was seen in a play, *Twilight Bar*, written by Arthur Koestler in 1933. In the play an enormous meteor flies over a town with a whistling sound and all the lights go out. In the play, the meteor is "a warning to the world." But this does not disprove the reality of the phenomenon: On the contrary, it suggests that something very real may

have communicated an actual warning to Stuart Whitman—
something that was simultaneously true to his inner life and
to the world around us all.

A 1950 novel, *The Flying Saucer* by Bernard Newman,
probably recorded the first instance of a flying disk having
an effect on a car ignition. Only after that date did reports
of flying disks killing car engines become common. But I
cannot assert that the disks don't *really* kill car engines, and
very possibly because of something about the mysterious
drives that power the ships.

In December 1985 a man had trouble with his car's electrical system after he had been taken by visitors. The car
lost all electrical power and could not be restarted even by
jumping the battery. It was towed to a repair shop where
nothing was found to be wrong. The battery recharged to
normal by itself during the night. I know that this story is
true because the man it happened to was me, and the problem manifested itself the morning after my December 26
experience.

Whomever or whatever the visitors are, their activities
go far beyond a mere study of mankind. They are involved
with us on very deep levels, playing in the band of dream,
weaving imagination and reality together until they begin to
seem what they probably are—different aspects of a single
continuum. To really begin to perceive the visitors ade-
quately it is going to be necessary to invent a new discipline
of vision, one that combines the mystic's freedom of imagination with the substantial intellectual rigor of the scientist.

There are many stories of visitors giving people secret
knowledge. Much of it has proved to be worthless or
worse, as was the information I got about electromagnetic
motors.

One fascinating case of transmission of knowledge dates
from the year 1491, when the Milanese mathematician
Jerome Cardan found himself involved in a visitor encounter. When he asked his visitors about the cause of the universe, "The tallest of them denied that God had made the
world from eternity. On the contrary, the other added that

God created it from moment to moment, so that should He desist for an instant the world would perish. . . ."

This is not a fifteenth-century idea. Upon contemplation, it emerges to a degree as a concept from Zen. More than that, though, it is a quantum-physical idea, suggesting that it is the observer who injects reality into the phenomenon observed.

Far from being a fifteenth-century idea, it is an eleventh-century Arabic idea. Did it get to Dr. Cardan via the sylphs, or was it transmitted by more conventional means? There is no other record of it in European scientific and religious literature of the period. If Dr. Cardan obtained the idea in a conventional manner, then what would have been his motive for risking his life to the Inquisition by admitting communion with the sylphs, who were viewed as demons? He was for his period an exceptionally rigorous thinker. He was sane and honest. A modern equivalent of what he did would be for a renowned physicist to announce that he had obtained important information in a UFO experience, but could offer no proof that the experience had happened. Perhaps Dr. Cardan was simply too honest to hide the truth.

In recent years many of the taken have reported having sexual experiences with the visitors. Currently among them this is a source of great disquiet, as will be observed from comments in the colloquy of the taken that follows this section.

It is terrifying, of course. But reflect also that mankind has had a sexual relationship with the fairies, the sylphs, the incubi, the succubi, and the denizens of the night from the very beginning of time. Nowadays men find themselves on examining tables in flying saucers with vacuum devices attached to their privates, while women must endure the very real agony of having their pregnancies disappear, a torment that I, as a man, doubt I can really imagine.

One of the women I know who has experienced this horror seems to me now like an anguished hawk, humiliated by her torment. I must add that no trivial explanation for what happened to her is viable: Her own gynecologist was deeply troubled and confused by the whole affair. To find some glib explanation for it is an insult not only to his skill and her veracity but also to her suffering, and to the worth of human suffering.

The Roman historian Suetonius maintained that Caesar Augustus was the product of relations between his mother and an incubus. Plato was also believed to be the issue of some sort of peculiar coupling, as was Merlin the magician, born of an incubus and one of the daughters of Charlemagne.

In a treatise called *On Little Demons, Incubi and Succubi*, written by Father Ludovicus Maria Sinistrari de Ameno in the latter part of the seventeenth century, some phenomenal affairs are recounted. One woman found herself awakened by a "fine voice, a high-pitched whistling sound." (I note in passing that when I first read this my blood went cold to remember that the thing that haunted me in Austin in 1967 made such a high-pitched sound. And in the *Malleus Maleficarum*, the notorious treatise against witches, demons are said to speak in reedy voices.)

The shadowy being with the high voice then announced its love for the lady. She was kissed so softly that it felt like cotton touching her cheeks. This went on night after night. The lady tried an exorcist, but to no avail. Eventually the incubus appeared as a boy with golden curls. Still, she retained her honor. He began to take her things, to strike her little poking blows, and to bother her in other ways. One night he built around her bed a wall of stones so high that she and her husband had to use a ladder to get out. I will let the good father recount the climactic assault of the lovelorn visitor:

"On the day of St. Stephen, the lady's husband had invited several military friends to dine with him. To honor his guests he had prepared a respectable dinner. While they were washing their hands according to the custom—hop!— suddenly the table vanished, along with the dishes, the cauldrons, the plates, and all the earthenware in the kitchen, the jugs, the bottles, the glasses, too. You can imagine the surprise, the amazement of the guests."

Indeed.

The teasing continued. In her efforts to rid herself of the incubus, the lady had taken to wearing a monk's robes. She was going to mass in the midst of a large crowd when all her clothes were suddenly whipped off her body and she was left naked in the middle of the throng. What could she do? She hurried home. Her torments continued, the good father relates with relish, "for many years."

Our present relation to the incubi and succubi may very well center on real flesh-and-blood visitors who first appeared here recently. But if so, then it also has a most profound and unexpected human dimension, for they are entering our consciousness where our gods and goblins live.

A famous case of sexual involvement with visitors took place in Brazil in 1957. The victim, Antonio Villas-Boas, had experienced a number of instances of strange lights appearing in his fields on the days prior to his experience. He was running his tractor one night when it died. Mr. Villas-Boas then saw that an object had landed in front of him. He was stripped, washed with a sponge, and taken inside the device. He was left lying on a table, naked. Sometime later he was astounded to see a naked woman, seemingly human, enter the room. Her hair was blond, parted in the center. Her face was extremely wide, her eyes blue and slanted. The face ended in a pointed chin. Her lips were very thin, nearly invisible. She was shorter than he was. Actually, she sounds very much like a cross between the individual I saw so clearly as the eidetic image and a human being . . . unless she was simply a visitor wearing what they thought would be a disguise. She made love to him, pointed to her belly and then to the sky, and left him. Later he was taken into a room with some males and attempted to steal a clock. As in dozens and dozens of tales in the fairy lore, he failed to get his artifact.

Why? Either because the visitors are good at keeping artifacts from us, or because they exist somewhere between reality and dream, and our inner selves know that we can never take into our hands things from the factory of the mind.

On December 29, 1980, a terrible event of some unknown kind took place near Huffman, Texas. (Oddly enough, this occurred on the same day and at about the same hour that a spectacular and controversial sighting was taking place halfway across the world in Rendlesam Forest, England.) A group of people observed a diamond-shaped object floating in the sky, glowing with a fierce light. The object was surrounded by helicopters. Some of these people were exposed to heat and apparently radiation. They brought suit in federal court, assuming that they had been the victims of a secret aircraft gone wrong. There was a lot of amused scoffing, of course, and the case drags on to this day. Meanwhile, these poor people have had their health shattered—a double mastectomy has been performed on one victim—and the government stonewalls.

The most interesting thing about all this material, the most important, haunting thing, is that in the past half-century it has slowly stripped itself of all the illusion, the armies in the sky, the fairies, the incubi, the glorious creatures of old, and come down to what it really is: a difficult experience, terribly enigmatic, the very existence of which implies that we very well may be something different from what we believe ourselves to be, on this earth for reasons that may not yet be known to us, the understanding of which will be an immense challenge.

Even the issue of where science stands in relation to this material has been with us forever. The first debunker was probably the Bishop Adelbard of Lyons, who in the time of Charlemagne saved from an enraged crowd three men and a woman who had been seen climbing down from an airship by half the citizens of the town. They claimed to have been taken for a period of days. The bishop saved them by announcing to the crowd that the whole thing was obviously impossible, and that people *had not seen* what they thought they had seen, nor had the poor victims been in any airship, because there were no airships. Thus the first debunker had the distinction of saving the lives of the first abductees.

People have not climbed down ropes from fairy ships since the turn of the century. Perhaps the parallel world has also had a technological revolution, or the mind of man has created new possibilities in its secret universe, or the dead have discovered wonders about which the living only dream. Maybe there really is another species living upon this earth, the fairies, the gnomes, the sylphs, vampires, goblins, who attach to reality along a different line than we do, but who know and love us as we do the wild things of the woods . . . who, perhaps, are trying to save us from ourselves, or whose lives are inextricably linked to our own. If we die, must the gods, the fairies, the elves then fall into some blue glen of unknowing? Will their secret world go cold without us, or will there only be less excitement?

If intelligence is normally centered in a hive or group context, a species such as mankind with individual independence of will might be a precious thing indeed, an almost inexhaustible reservoir of new thoughts and ways of acting.

Up to a point, there would be a tendency for the hive minds to isolate us, both to protect our freshness and to protect themselves from us. But then, as we matured and came to understand them more clearly, the potential to enter into a relationship with us would emerge. For such a species, old and with its single enormous mind essentially alone, that potential might eventually overwhelm even the most rigid instinct to self-preservation, especially if we were to learn a way of approaching them that would not threaten them.

This thought leads inescapably to the issue of modern abductions and encounters. They seem qualitatively more "real" than those of the past, although the extended visitations experienced in France, Japan, and other places in earlier times also imply wide contact.

In *An Essay on Man*, Alexander Pope said the following:

> So man, who here seems principal alone
> Perhaps acts second to some sphere unknown.
> Touches some wheel, or verges to some goal,
> 'Tis but a part we see, and not a whole.

The Hidden Choir

Budd Hopkins has developed great sensitivity to the problems people face after they encounter the visitors. He has dealt with more than a hundred cases, and knows the pattern of response. When he suggested that I meet the loose support group of others in the New York area, I was at first relieved. Then I became uneasy. "Don't worry," he said, "everybody half believes that they're dreaming all of this up. And that's the healthiest way. Nobody is going to show you an extraterrestrial belt buckle and blow your mind."

Still, I was not eager to meet the other "abductees." Just a few days before, I had interviewed a person who believed that he had been contacted by people who "gosh, just looked like the most beautiful gods and goddesses you ever saw," who explained to him that the world was soon going to end and that the "chosen" would be taken to live on a moon of Jupiter. I hope it isn't Io. This man described a familiar initial visitation, but had altered the terrifying and uncontrollable parts into a structure of belief congenial to him.

I expected to encounter people who hungered for belonging, for publicity, who tended to the imaginative and the grandiose, and who were a bit paranoid. I anticipated that their psychological deficiencies would be obvious to me.

This was all very far off the mark. They wanted nothing to do with publicity. They demanded anonymity. They were a group of average people. I cannot seriously maintain arguments that they are insane, or even particularly unbalanced. They were all anxious, that was obvious. Under the circumstances any other reaction would have been abnormal.

The group was for the most part rather hardheaded and not unusually imaginative. Among them were a business executive, a cosmetologist, a scientist, a hairdresser, a former museum curator, a musician, a dancer—in short a cross section of any big city. They clung firmly to the idea that they might have been dreaming, clung to it, I thought, as to a bit of driftwood in a storm.

I found that my experience had many similarities to those of the support group. We have almost all seen versions of the same creatures. Some of these are small and quick, wearing gray or blue uniforms. Others are taller, graceful, and thin, some with almond eyes and others with round eyes. I have also seen, in my childhood, a very commanding presence in white, which had light blue eyes and skin as white as a sheet. This came back to me in the form of disjointed memories apparently dislodged by all the thinking I had been doing about this subject.

Other relatively common observations are the seemingly ubiquitous gray table with the solid base, the smallness of the visitors, their large, black eyes devoid of iris or pupil, and the fact that there is either more than one type or more than one species appearing in the same context. Many of us also seemed to have relationships with particular beings.

Their skin tone seems to be gray, with other overtones. When they speak aloud, it is sometimes with a high, squeaking sound, other times in a deep bass. They can also create words inside the center of their heads. One occasionally feels from them powerful emotions. Other times they are as emotionless as stones. People report various smells, primarily pungent. Light, both as a means of anesthesia and as a medium of transport, is commonly described. "I rose up the shaft of light" and "The light hit me and I was totally paralyzed" are typical statements. Electromagnetic effects are also commonly reported, primarily malfunctioning cars, television sets, and home lights.

A number of us have also been in a small operating theater, but nobody seems to remember what transpires there. One woman was left to walk around in such a place by herself.

Interestingly, one sound that is reported other than the various voices is a very low-pitched noise. There is a small body of research suggesting that low-frequency sound may have biological effects, especially in the area of disorientation.

There is as well a striking symbolic consistency, which lies hidden within many of the accounts I have heard and read. It has almost no reference to modern Western culture, and so is not particularly likely to have been drawn from some general pool of background symbols.

But the symbol is very ancient, as it happens, and through much of human history was tremendously important. I have had a lifelong interest in it—really, an obsession. The others in the colloquy all noted its presence. It is mentioned in many of the tapes people have allowed me to hear, and it appears in the drawings they have made. It is an incidental, though. Before now, nobody has seen it as a general symbol of the visitors.

This symbol is the triangle. Buckminster Fuller, in his autobiography, called it the "fundamental building block of the universe." It is the central symbol of growth in many ancient traditions. An understanding of it is the key to the riddle of the Sphinx and to the pyramid as the mark of eternal life. G. I. Gurdjieff relates it to the "three holy forces" of creation and it is the main sense of the Holy Trinity.

I had a pair of triangles etched on my arm in February 1986. "Dr. X," a physician in Arles, France, who prefers to remain anonymous, had a triangular rash appear around his navel after his experience.

Sifting through this colloquy will be the symbol of the triangle. While the colloquy was taking place neither I nor any of the participants was aware of the symbol's importance.

When we have contemplated sending a message into space, we have thought to send some core symbol—a prime number, perhaps, or the value of pi. The transmission of an isosceles triangle would not be an invalid choice.

On the night of April 13, 1986, eleven of us met at the home of Budd Hopkins. We were selected simply on the basis of the fact that we live in the New York area and could come.

During the colloquy I persistently asked that specific experiences be recounted, but did not have too much success. To many of these people, the details of what happened are an extremely private matter. And given the shrillness of the debunkers eager to accuse them of everything from charlatanism to insanity, and elements of the press so eager to scoff, I could not really blame them.

The purpose of the colloquy was not primarily to discuss the details of being taken, but rather the experience of coping with it, of trying to live a normal life without knowing for certain what is real, of facing the risk of personal and public ridicule, of finding one's way in a world that has suddenly become very strange indeed.

Needless to say, none of these people would allow his or her name to be used. The only real names in the colloquy are thus my own and that of Budd Hopkins.

This is our hidden choir:

> Mary, cosmetologist, age 29
> Jenny, dancer, age 22
> Mark, museum curator and artist, age 55
> Sally, business executive, age 36
> Joan, beautician, age 23
> Sam, scientist, age 39
> Fred, musician, age 34
> Pat, housewife, age 35
> Amy, Pat's mother, age 56
> Betty, executive, age 43
> Whitley, writer, age 40

This is our colloquy.

Whitley: "Budd, I'd like it if you could begin. Even though you aren't one, you're still one of us."

Budd: "I'll tell you what I think would be the most interesting thing—rather than tell their experience, why not focus on the idea of how everybody feels about their experience?"

Whitley: "But say what happened to you so that there'll be some perspective in people's minds when they read it."

Budd: "The most valuable thing, really, is for everybody to say how you handle this, how you fit it into the rest of your life if you do, and how seriously you take it, and how important it seems to be to you and so forth. That's very crucial."

Joan: "Sometimes I have a problem feeling the importance of what's going on now, as far as things that take place in the world and jobwise, and the whole attraction of life itself, because I start thinking that this is so mediocre compared to what's out there. What we're doing—people put so much attention and so much pressure on whatever they're doing in their lives, sometimes it gets to seem like we're such jerks, and I say to myself, 'It doesn't mean anything.' There's something that's gonna happen soon, and this doesn't mean anything, what we're doing. And they're trying to tell us something, but nobody's listening."

Whitley: "What happened to you?"

Joan: "I'll tell you one thing. I was shown a picture of another city they are building. What we're doing now to our planet is killing it little by little, and it's going to come to a point where there's not going to be anything left. I think that they're getting ready to start another world. And there will be people who are a part of that. And it scares me, because I have trouble dealing with what's going on in my life now because I start thinking, *This isn't really what's happening*. It is ending, and they're telling us that, and they've implied that to me. What we are doing is killing ourselves. And that's scary."

Whitley: "Any other thoughts?"

Jenny: "I think what she's saying in terms of the mediocrity of what we're going through is only in the eyes of people around us, but that the important thing is right here, and some of us really understand what is going on, and maybe they are not 'them,' but they are us and we are them, so if you call them 'them,' and say, 'They are looking at us, they are doing this to us,' it's not right. They are us and we are them, and so . . ."

Whitley: "What happened to you?"

Jenny: "I'm not really sure yet because I've only had one hypnosis, but I remembered something from when I was five years old, a very scary experience, and I've always blocked it. From the time I was five I was afraid. And I saw things in my house, I saw people in my house, and I would wake up screaming."

Whitley: "You mean, not human people?"

Jenny: "I don't know, they were shadows. Small things. I saw once this green thing dripping down the wall. It looked like a very bright green triangular light. And I went screaming into my mother's bedroom, and she said, 'Just go to sleep. Obviously a dream.' And so those are the kinds of things I saw from the time I was about five, and I never connected it with anything, until about six months ago my sister said something to me about it, an experience that she had that she remembered me being in, and I remembered it but I'd thought it was a dream."

Whitley: "In February I had a triangular piece taken out of my skin on my arm."

Mary: "The best way for me to live with this is just not to believe it. I mean, there's a part of me that doesn't. The part of me that lives every day doesn't."

Whitley: "How much of this experience have you had?"

Mary: "A lot, since I was about five."

Whitley: "How much, would you say? How many times?"

Mary: "Seven. Eight, nine, ten."

Whitley: "Has anything happened to anybody else you know?"

Mary: "My whole family. Neighbors, quite a few friends. From before I ever knew them. We've all just come together. Several generations."

Budd: "You said that there was one figure, one man—"

Mary: "There's always been one central figure."

Budd: "And he was nice?"

Joan: "Was he tall?"

Mary: "No, they were all little guys. He was my protector. Everybody else who was around was always really very—they were doing a job they needed to do, and that was it. There was no—they weren't angry or mad or happy about it. They were just doing what they had to do. But this one guy, in all instances, this one guy—when I got

254

scared he calmed me down, when I felt bad he made me feel better." (Note: Others have had a very similar experience of a "friend" or a "protector." The perceived sex of the guide is not consistently opposite, but very often is.)

Whitley: "What does he look like?"

Mary: "He looked like all the rest, really."

Whitley: "Which is?"

Mary: "The same small people, you know, four and a half or five feet tall. With the gray skin and the large heads and the big, fluidy, black eyes that went on forever."

Whitley: "In other words, you wouldn't identify him as a human being."

Mary: "No, not in my town, anyway. New York is a different story. Regardless of whether or not any of it was anything more than a dream or what, I know the emotions I've had to deal with through the years have been real, all the anxiety that doesn't have any source."

Whitley: "Any of your kids?"

Mary: "Yeah."

Whitley: "How do you feel about that?"

Mary: "That's the only part of it that really upsets me. I guess it's just the mother instinct in me. I want to know everything that's happening to my children and I want to be there when it's happening. I don't like the idea of someone screwing around with my kids, and me not knowing about it. They're so defenseless, you know, it's not fair. As if we aren't all as defenseless as children, but somehow my mothering instincts are the only thing that turn me on, make me mad. There's only one thing that I'm really angry about, and I don't really understand that. I told Pat, I don't feel like being angry for me, anyway, is going to work. I have enough trouble dealing with everyday life. There are a lot of stresses, and I don't cope with stress really excellently well. I just get by, like everybody else. So I don't see any point in inflicting any more stress on myself by getting all worked up and angry over something I have no control over. Something that no matter how damn mad I get, it ain't gonna do no good. I'm just gonna make myself get worse. So I haven't whipped up a lot of anger over this, other than about my kids."

Whitley: "So basically, you just decided—"

Mary: "Live with it."

Whitley: "Keep it from getting too real?"

Mary: "I believe something is happening to everybody. At least everybody here. I don't know why it is more easy to accept things from all of you than I find it to accept them for myself. Like saying it seems more real when it happens to somebody else."

Fred: "I just want to say that I think we're dealing with so many not-so-obvious things. At least one thing which I think is obvious and important is the fact that we meet. I think that's important. If there's anything we can understand, it's the meeting. The experience—I speak for myself, I don't know how traumatic it was for others—it was fine for me. It was mind-opening for me. It's thanks to this little group that I'm getting to know—love—that's what's important. The rest is up for grabs."

Whitley: "You want to say anything about what happened to you?"

Fred: "Not really."

Budd: "When people first come, they say, 'I feel it's like family.'"

Whitley: "That's what I feel. It's very strange." (I was privately contending with the fact that the one called Mary in the colloquy was instantly recognizable to me, and I did not know why.)

Mary: "Strange for me, too."

Pat: "I think the most interesting thing is that I went to a meeting up in Massachusetts and there was a whole bunch of people—and for some reason three people who had been abducted all came together in the middle of the room. It was very strange, and we all knew immediately. And there was nobody else in the room who had any experience—"

Budd: "Don't count on it."

Pat: "I'm not saying that. What's interesting is that they knew. They found each other immediately. We huddled around each other. It was almost like we needed to be together. And it was very strange."

Betty: "I think you get to the point where you almost want to detach yourself from the situation, because you'll really just lose your mind. You have to look at yourself like an outside observer. That's just not happening to you."

Sally: "I think that's the reason it's so difficult to piece it together when it's happening, because while it's happening you're saying this is not happening to me, so part of the mind shuts it off and part of the mind is having to absorb it. So I think that's the reason it's so confusing when it's happening. At least that's what I felt. I felt, really, that my whole mind was just falling apart. It was just crumbling."

Whitley: "What was happening to you? Can you describe anything at all?"

Sally: "Absolute terror. I felt like an animal, totally warped and totally working on the instant."

Whitley: "That's how I felt."

Sally: "I was just clinging to any little piece, little scrap of life. Any kind of shelter, if I could hide in a corner, if I could get away from them somehow. I didn't want to know what they were about, what they were going to do, what they wanted to do with me—I just wanted out. Get me home. That's it. I make no claims to being brave. I was not brave. I said often that I felt like Fay Wray. I was screaming and passing out. I don't care about evolution, I don't care about your spaceships, I don't care about anything. Just let me out of here. Of course there were times when I was less frightened and I looked around. But when they were there, no. Most of the time I was angry or terrified. That was it."

Fred: "I still am surprised, despite all the books I've read, I find more out here at this group that makes sense to me personally than I do from any books."

Whitley: "Sam, you're just sitting over there—"

Sam: "My experience happened to me a couple of years ago. I guess it isn't much different than anybody else's." (He had a particularly startling experience, especially for a scientist.) "I find it's easier to sit here and talk to other people about it and listen to other people's experiences than to sit in a quiet corner and get into my mind, and what happened. It's very difficult, almost impossible, for me to close my eyes and go through the experience. I can't do it."

Sally: "It's too scary. It's much too scary alone. I did a self-hypnosis thing actually, I—"

Fred: "My God, you've got guts!"

Sally: "Well, I did. I went to sleep. I started to relive the actual abduction. I was in my apartment building going up the stairs. Then I got past a certain point and I said, 'Oh, no, this is too real.' 'Cause I was actually remembering more details that I had under the actual hypnosis. I said, 'Oh, no, this is not going to be able to work.' I stopped."

Sam: "When I get to thinking about it alone, by myself, I get a little angry, and I begin to think, *Who the hell do they think they are that they can just do what they want to us, as though we're nothing.* And that really disturbs me, so I turn it off, and I don't want to look at my own experience, and I don't even want to think about what's happened to others, because that disturbs me, too."

Budd: "Mark, if you have anything you'd like to say at this juncture. You've gone up and down about your feelings about it and how real it was. Curious to hear you talk about that."

Mark: "Just trying to get a little understanding. I had an experience when I was ten years old, had no idea what it was, but I know for thirty-seven, thirty-eight years, I was always aware that something had happened, and a general idea of the location. But I could never explain it. I constantly looked for the area where it had taken place. It was in an area I went through often, where there were a lot of people that are witness to the fact that there was something you just couldn't explain. It was after the first hypnosis that all of this comes out, too real, too believable. I was with another person, we were out bike riding. Then there is a lapse of time and we're walking our bikes home. I remember telling a story that I'd had an accident on my bike because I had a scar, but not believing myself. And not believing it throughout the years."

Whitley: "I'm very curious, just to interrupt for a second. How many of you have stories like that, about things you know didn't happen that you've been telling all your life?"

Jenny: "All my life I used to hear my mother—in my head there was this thing saying, 'You're lying, you're lying,' but my mother never said it. My whole life from the time I was five years old."

Fred: "Yeah, yes to that question from my point of view also. I know it happened but I can't believe it. And the other thing is, what's nice about the group, to get back on that for a second, we look at each other and we say, 'I can't believe her story, I can't believe his story, I can't believe your story, I can't believe *my* story.' And yet, there's a comfort which we still share because behind it all, oh, it's all the same. We don't understand it, but something happened."

Pat: "There's an acceptance."

Sally: "You know, the first time I realized something had happened that wasn't a figment of my imagination, wasn't some subconscious thing or something or a creative element of my mind or something, I was reading Betty Andreasson's story [*The Andreasson Affair* by Raymond E. Fowler]; she described a detail of some sort of crystal boots they had put on me. They had clear bottoms, like a platform, and there was some sort of electrical thing, something inside this clear platform. And that's exactly what I had on my feet. Exactly. And I said I cannot believe some other woman could possibly have had the exact same thing on her feet, that her imagination could be exactly like mine. And I just said, 'No way,' and the tears starting coming down my face, and I said, 'That's it.' And I was totally upset. I couldn't sleep, I tossed and turned, I was just a mess. I wanted to hide somewhere. Horrible."

Joan: "Isn't it a relief when you find it isn't your imagination?"

Sally: "No! Horrible! Except if I was called a liar. But otherwise, no."

Joan: "I figure it's a relief."

Whitley: "I would have gone insane if I thought this was

my imagination. At first it was perfectly obvious to me, I was going crazy. I expected to just go around the bend. The realization that it wasn't my imagination, when they came in such a way that I couldn't deny it, even if I wanted to—"

Joan: "Has everybody had an experience when they were five?"

(General agreement. Some said maybe four, or very young.)

Whitley (to Amy): "Any thoughts? Do you know what happened to you?"

Amy: "Yeah, I know. I want to say what Mary talked about. One time it seems real, and the next time it's not real."

Mary: "Every time I look at the pictures of my backyard, then it's real." (It was from her yard that Budd Hopkins obtained the sample of calcinated earth.)

Amy: "Sally mentioned Betty Andreasson's book. I looked at a few pages and I couldn't read the book. I knew I would be terrified and I don't know why. I had this trouble feeling like—I have daughters—and I was afraid."

Tom (a college teacher who does UFO research but has not had an experience): "In a way I feel envious of everyone here because you've all had a glimpse into another world, another dimension perhaps. And in a way you've seen the future, if I could even say that, which may or may not be true, if I could even say that. You have seen what might, in fact, be coming eventually down the line. At least, there are people who believe that. And so, you all have a sort of special knowledge that very few other people have."

Sally: "The question is, what kind of knowledge is it and what if we don't really want it? And if you don't want it, then you refuse to accept it, and if you refuse to accept it, you don't have it, so it's sort of like this whole thing—it's like if you could see something inside of a ship and say 'Is that real?' because you've never seen it before. We're not really good witnesses. I mean, we have just little pieces of things. If we were just allowed to explore one of these ships, imagine the information we could get. But just with little innocent people being abducted, it's not enough. Even though it may be a glimpse. It's a fragment."

Whitley: "I don't think it'll ever come out completely into the public eye. And when it does, it won't be as intimate as these experiences. People will see it, you know, like something in the sky that everybody sees and it's there for four days, that kind of thing."

Sam: "At what level of belief are we? Do we all believe that we had single experiences and they're gone? Or do you believe they come back and forth type of thing? Do you believe they're here all the time among us?"

Pat: "How many people have the sense of continuous monitoring?"

Jenny: "Being watched all the time?"

Pat: "Continuous monitoring."

Joan: "I have a very strong feeling."

Whitley: "I do, too."

Pat: "How many people have the sense that there is something involving permanent relocation?"

(Mixed reaction.)

Whitley: "I have persistent images of being in another place. Sometimes it's parklike, sometimes very bright."

Fred: "I do, too. Very bright."

Pat: "What makes us afraid of the change?"

Joan: "The sense of not being in control."

Pat: "I don't think they have individual DNA. I think they're all pretty much the same. They're interested in us because we are different. And we value the difference, and our individual freedom. And we feel that when we were abducted, that individual freedom has been taken away, and they don't understand that. They don't really understand our sense of freedom and being allowed our own will."

Sam: "They were almost like under military discipline."

Whitley: "That was my impression."

Sam: "They had instructions and they followed their instructions, and that was it."

Whitley: "Do you suppose we see robots?"

Sam: "I thought of that."

Mark: "Fanatic, or just disciplined?"

Sam: "Disciplined."

Jenny: "That's right, I remember them all walking the same way."

Sally: "Moving in unison, speaking in unison?" (Unified movement is often reported, such as three individuals walking in lockstep.)

Joan: "I can imagine something above them that's speaking through them. They're to do their job."

Jenny: "You don't feel any personality—"

Whitley: "And yet at times I feel incredible personality. She's the strongest personality in my life."

Jenny: "Oh, I feel that way too! I feel a personality but I don't know where it comes from. When I try to picture them, they are one sort of thing, going. But there's this sort of force—"

Pat: "One that cares a great deal."

Fred: "There is one, at least in my case, that I can identify that had that force of personality. He was directing the whole operation. Everything. The others were simply not even taking orders. It's not that an order was barked at them or said to them. They just—boom, boom, boom, did it. I felt like I was even superior to them. I felt like slapping one of them. There was always one you felt comfort under, security under."

Jenny: "Like he was part of me. In me."

Mary: "There's a little part of him in me all the time."

Sally: "One did try to comfort me but I refused it. I didn't want to buy any of it. Actually, what my feeling was, I felt that if I looked at him, I was—sort of like looking at the person who robs you. Someone comes up and he has a mask on and if you take down the mask, uh-oh, you're going to be killed now. You've identified him. So I didn't want to look at him. I didn't want to believe it and I didn't want to become a part of it. I felt if I removed myself I would be safe. I won't look at you, I won't identify you, I won't tell anyone what you look like. So I just said, 'I won't look at you.' That was just my feeling. If I could identify him my life would be in danger." (Note: Sally's experience involves seeing not only visitors but seeming humans involved with them. It is these people that she apparently did not wish to recognize.)

Sam: "It felt like one superior intelligence. Very, very powerful sense of intelligence. All the rest were nothing."

Sally: "The others weren't human like that one. The emotions— The one that speaks to me is a mix. He's a crossbreed between them and ourselves."

Whitley: "The one that speaks to me looks like a big bug. Big eyes. Doesn't look anything like a human being."

Sally: "Yeah, I saw those."

Fred: "Just a question. How many feel used?"

Sally: "Yeah."

Sam: "It could be harmless. Harmless use."

Mary: "You know, like for a whole year I was obsessed with taking some little piece of the world with me. I took my children to the park and I collected every little seed and rock and twig I could find. My whole room looked like a nature study. Then half of the stuff turned up missing."

Whitley: "You know, I sometimes have a thought of what will I take with me."

Mary: "I do too. I keep thinking that one day this will all be gone. But my thinking is automatic: I want to have this so my children will know how it used to be."

Sam: "I feel the same way. I want to enjoy what's left here, because it won't be here much longer. I want people to understand that."

Mary: "My sister was left with a thought twenty years ago, and she still believes it strongly, that by the year 2000 this world is going to be a totally different place. It won't necessarily be a bad place for a lot of 'em. But for some of 'em who can't adapt, survival will be very difficult. But it will be a good place. The world will be a more stable place or maybe a different place."

Sam: "More artificial."

Mary: "Truly different from this place here. That's for sure. She says that it is for the young and strong. She doesn't quite know whether that means physically young and strong or mentally young and strong. But it will only be for those people."

Sam: "In fifteen years?"

Whitley: "I have a feeling it's right on top of us, too. My feeling is that a cycle sped up recently, a lot. It's going fast, not slow." (I wish to add an aside here to expand on why I made that statement. In the form of what can only be described as vivid bursts of information, I have received a great deal of material about the perilous condition of the earth's atmosphere. Much of this material came in February and March of 1986 and concerned the danger of impending atmospheric deterioration. In March I called a press conference of environmental reporters in Washington, D.C., to discuss a book I wrote with James Kunetka, *Nature's End*, and to warn about the serious implications of the hole that had been detected in the ozone layer over the Antarctic. My apparent visitor information suggested this hole would lead to further holes over the Arctic and thereafter a thinning of the ozone layer over the Northern Hemisphere, with measurable crop damage from excessive ultraviolet light beginning to occur in the 1990–1993 period. At the time I gave this warning, the only stories about the hole were saying that its significance was not understood. I had also been told that the atmospheric problems will cause a reduction in immune system vitality in all animals, and the consequent resurgence of disease. As yet I have seen no scientific corroboration of this. The damage to the immune system seemed related to excessive ultraviolet light, but the information emerged visually, appearing as complicated images that I may not have understood fully. This information may indicate, by the way, that the ozone holes will open and persist in a single place rather than circulate. Additional information suggested that volcanic activity had exacerbated the ozone problem, and that there would be some reduction of the size of the major holes during the eleven years from 1986 to 1997, but that this would be only a temporary respite.

After the conference I was careful to mention to one reporter that I had had an extremely strange experience that had, in part, led up to my warning. I told this to UPI reporter Ed Lion and explained that it would be the subject of my next book. Lion asked what exactly I was talking about. In his May 16, 1986, review of *Nature's End* he wrote, "Asked what that subject could possibly be, he shook his head mysteriously. 'You'll have to wait.'" I felt embarrassed to have been so cryptic with him, but I wanted a record of some sort of the real motive behind my calling the conference. If I had told the reporters that alleged visitors were in any way involved, it seemed to me that there was a high probability that my credibility would be destroyed. I hate the idea of repeating predictions because I cannot assess the correctness of the "information" that appears to have come to me through the visitor experience. However, in the past year the atmospheric predictions have emerged as being quite accurate, and I thought it best to mention them in view of the obvious seriousness of the problem that they address.)

Sally: "How many people really feel used? Is that a feeling?"

Mary: "I don't feel it's bad."

Amy: "I wouldn't use that adjective. Things happen in certain places to certain people."

Sally: "I felt somehow psychologically violated. There was nothing in my life that would indicate that kind of trauma. There was nothing in my life. Then what happened when this media thing happened and they used my name I felt like it was happening all over again. Horrible, horrible! I couldn't control my crying, couldn't control this horrible feeling inside myself. I felt so powerless! It came back at me and I couldn't look at anything that said *UFO*. I couldn't stand it. It was just tearing me up inside. I said, 'Something must have happened.' It had to have happened, because this feeling was so intense. It was too intense to— True, you're gonna get angry because they're using your name, but there was this absolute gut-wrenching feeling. It was just awful. I said, 'It's happening all over again. Somehow or some way,

I'm being stabbed in the back again.' And it was really horrible. It was just a horrible feeling. And that's what I mean by used." (Sally's real name was discovered by some journalists who held her up to public ridicule.)

Amy: "There's something about Betty Andreasson's book that's awakening a terror in me. Something that I've put away, that it was bringing out. That's why I couldn't continue to read it. If I continued to read it, it was going to come back to me and I don't want to know. Whatever that means."

Sam: "Does anybody ever experience light without any source? You see it on the wall or on the ceiling. It could be in a triangular shape or round. Sometimes I see a triangle. Three triangles together on the ceiling. Has anybody else seen that?"

Mary: "My son saw a red light chase him out of his room the other night. He called it a red tarantula. It was a small red ball of light with things sticking out of it. That was the same night that one of the few things I believe *really* happened, happened to me."

Whitley: "I saw a light just a couple of weeks ago that went down the hall and into my son's bedroom. I ran in there, but I couldn't see anything wrong."

Sally: "That's what I saw coming down from the roof. Then the whole area in the hallway was lit. Totally lit. Then what happened, I was starting to go up the stairs and I turned around and saw my shadow cast from the light. Then the light went out, and I decided to go up to see what was going on."

Sam: "Has anybody had their TV go on or off by itself?"

(General reaction. My own television went off at the main switch a few nights earlier, so that it had to be reset before the remote control could be used again. My wife, myself, and a scientist friend have observed that I can sometimes affect electronic devices by simply placing my hand near them. Since the energies that activate such instruments are known, we intend to devise some physical tests to attempt to measure this. The problem, of course, is that there may be other energies we know little or nothing about that could affect electronic equipment.)

Mary: "Before any of the stuff in the yard, either right before or right after the incident of the three men in my room who gave me the box—"

Whitley: "What was in the box?"

Mary: "I don't know. They just told me to look at it and said I would remember what it was for and how to use it when I saw it again. And I was to hold it and look into it, and I did. But my TV would turn itself up and down all the time. It would turn itself on in the middle of the night. So we finally unplugged it."

Whitley: "I walk into rooms and short out stereos and things to the point where there are people who won't let me get near their sets, and get mad if I touch the equipment because they say it shortens its life."

Jenny: "Ever wake up and the blue thing is going like this?" (Makes pulsating motions with fingers.)

Whitley: "What blue thing?"

Jenny: "The blue TV light."

Whitley: "There is no blue TV light."

Jenny: "What?"

Fred: "I would turn the TV off, go and sit down and start reading, and the TV would go back on."

Sally: "I got the idea that I could stop the roll over on my TV set. I would generate energy out of my hands. This was after the abduction. I don't know where I got the idea I could do it."

Whitley: "Is there anyone here who doesn't do things to electronic stuff?"

Mary: "I put my hand on a TV screen once, and the TV was turned off, and it had been off a while, and when I took my hand away I could see its outline."

Whitley: "I think everyone can do that."

Jenny: "Sometimes I wake up in the night and it's blue but there's no picture."

Sam: "A strong surge of current crosses the switch . . ."

Mark: "I think I've had two happenings, one when I was about ten and one about fourteen years ago when I was teaching. That event I—it was very strange. It happened in a place where I thought it couldn't possibly happen and anyone not see this happen. I just sort of passed it off as a story I sort of made up. But I don't understand why I went and told the story. And then I'd also go back and try to find where this happened. I was driving out one night with the dog, to go and walk the dog—which isn't something I would normally do, to put the dog in the car and *drive* somewhere to walk the dog—and I drove past this area and I saw this light. It pulled the car over. Or I thought it pulled the car over. I don't remember getting out of the car, but I remember somehow getting to the area where this was taking place, and I see this silhouette shape and this light source where these little men, these little creatures, come out. One of the three sort of comes close to me, but nothing is ever said, nothing is done. Next thing I know I got back in the car. Now, the area it happened in, it's totally impossible for something like that to happen in, without people around knowing it happened. My question to you is, have you had these things happen in places where it seems totally impossible to happen?"

Sally: "Mine happened in the Bronx. It's in the middle of New York City, but it's a quiet area. We think we had some witnesses, but we don't know if these people still live there. I'm sure somebody has seen it. But nobody told anybody. But there it was on the roof. It was as Whitley once described, a dark shape, you couldn't see the sky through it. But there it was in the middle of the Bronx. There were people down there and there were cars and lights."

Mark: "When I think about it, I just don't know if it was me that stopped the car."

Fred: "They could have taken you somewhere and given you the impression you were in a populated area."

Budd: "Mark said he was walking his dog in this park. So I naturally made the assumption that he just went out the front door and walked the dog. So, in other words, it was close. He said, 'No, it was quite a distance.' So I said, 'You really walk your dog a long way?' He said, 'Oh, no, we went by car.' I said, 'You walk your dog by car? Do you often do that?' He said no. I asked if he'd ever done that before. He said no. Then it began to look very odd to him— 'What in the world was I doing walking my dog by car?'"

Mark: "I haven't read Budd's book [*Missing Time*]. I got a hundred pages in and it began to seem too familiar. I didn't want to read the rest of the book because I didn't want to be influenced by anything I read that wouldn't be of my experience, during the whole hypnosis thing. I've got to keep this as pure as I can."

Budd: "When Mark's lady was present when he described this place he was when he was ten, which turned out under hypnosis not to be a place but a thing, more or less, she said with great relief, 'I've spent seven years with him looking for that place. He's got me believing it was a real place, a tunnel. Thank God I don't have to look for it anymore. The mystery's solved.' You were so convinced that it was a tunnel." (Note: Both of Mark's experiences involved extreme disorientation as to place. This is quite common, but appears to have confused him more profoundly than most.)

Mark: "About a year ago I asked my mother if she remembered that incident, and what did I tell her. She remembered it, and repeated what I had told her. When I asked her where it had happened, she said it had happened at the end of the street. She went down and was relieved I wasn't killed. There's a slope. It's a hill. And there's a tunnel there wide enough for four lanes of traffic. But when it happened it was like I was far, far away from home, on the other side of town or something."

(I then asked the group what their jobs were. There was revealed a profile of people on the run, constantly making changes, moving, leaving, escaping. One of them, Jenny, had just fulfilled her lifelong dream to move to New York and "live in a big city full of lights and lots of people." She has had this aspiration since she was nine.

I grew up with the same dream, to live in a little apartment in an enormous building in New York with a view of a brick wall.

I did that, though, and it didn't help at all. I now live in an apartment with huge windows and spend a lot of time in a very isolated country cabin.

For most of my life I was running from this, whatever it is. I am unwilling to run anymore.)

Whitley: "We all have trouble saying what we did."

Joan: "Why do you think that is?"

Whitley: "I think we all have something called performance anxiety. That's one of the reasons we all have difficulty pinning ourselves down to do something. I've been extensively psychologically tested twice, and in both cases it came out that there was this deep anxiety in the performance area. I think the reason is that a lot of us have been asked to do some very hard things, that were very frightening, and also I have a feeling that we've been questioned very, very hard. And we can't remember that either."

Jenny: "Do you have trouble taking tests?

(General agreement.)

Whitley: "That's performance anxiety."

Jenny: "My profession involves auditioning. And every time I go to an audition, I think I'm gonna die."

Sam: "I have tremendous anxiety. Things that I'm very sure of, in the sciences, I don't have a difficult problem. When you start getting out of those areas, I get into real problems."

Whitley: "Fear?"

Sam: "I don't know what it is. Some kind of anxiety there. I really can't pinpoint it."

Sally: "The simplest test—a typing test—I freak out."

Sam: "Being examined . . ."

Budd: "You see, if you imagine that there's a situation of having seen two worlds, of living in this world and then being dumped into this other world at intervals—that has to make you wonder where you belong. And if, in that other world, you are deprived of your ability to act on your own— you can't even move, and you have no choice, nobody asks you anything—you doubt your own powers. In a certain sense you are physically impotent, unable to do anything."

Sam: "And when we're actually being tested, the anxiety is enhanced."

Budd: "One very strange thing that Pat remembered—the thing of the needle going in under the eye—another case of the needle going up the nose. The neurosurgeon said that is going into the region of the optic nerve, and he said, 'Wouldn't it be wild if you could see through people's eyes?'"

Whitley: "What if I say—love? Longing? Does anyone feel anything like that toward them?"

(There was general agreement, except for Sally, who demurred in this way:)

Sally: "You know, the funny thing is, before hypnosis I felt an attachment to them, a love. And after hypnosis I was angry. So I can't really say that I feel love."

Pat: "I feel loyalty."

Sally: "I feel like I want to strangle them."

Sam: "I have a lot of mixed emotions. Why are they doing this? That aggravates me. I go back and forth because I don't know. So it's mixed feelings. I feel that they demand loyalty."

Whitley: "I think we're sisters and brothers not from the fact that we went through something together but from the fact that we noticed."

Sally: "What I really feel hurt about is that I wasn't given more respect. If I ever meet them again, I want to be in control. I want to be able to speak my mind without them telling me what to do. I want to ask them questions. I want some respect for my being. And if they don't do that, then I don't want them back, I really don't. I don't want them anywhere near me, if I don't get that response."

Amy: "You want them to share with your intelligence what they're doing, rather than forcing you to be a part of it."

Sally: "Yeah. I don't even want to know what they're doing. I may be curious, but the point is, if they can't trust me, why should I even care about what they're doing? I just don't want to know. The human race has to have some sort of respect from these creatures, and we're not getting that. I don't feel it, and I'm certain—they take children in the middle of the night, they don't care about the anxiety of the parents—they don't understand so many things, and they don't make an effort to understand. Until we can get across to them that we matter and deserve respect, I don't think we should give them respect."

(More than one participant was aware that the visitors were involved with their children. One man, who saw his child taken in the middle of the night while he was himself entirely conscious, disagreed categorically with Sally. He maintained that they had shown him respect by allowing him to know what was happening to the child, and not only that, that the child was unharmed in the morning, and seemed filled with a new light of the mind. That day, the child made the following comments: "Reality is God's dream," and "The unconscious is like the universe beyond the quasars. It's a place we want to go to find out what's there." He also said, "Dad, I had a dream last night. It was like a dream but it wasn't a dream. I was in the woods and a huge eye was looking down at me.")

Sally: "I'd think you'd feel horrible about it."

(The man added that he did feel horrible about it, but at the same time felt that they took the child because they had to or needed to very badly, and they had been helpful and supportive of him while the child was gone. "Then I woke up in the morning and my child was fine. More than fine. That's the reality.")

Sam: "Do you think they possibly match emotions? If you are hostile, they will be hostile? If you are not forceful in any way, they will be nicer? If you comply, they will, too?"

Whitley: "They won't comply, I don't think. They can be nicer, I've seen that."

Sam: "They're never really hostile with children."

Whitley: "My son remembers them saying in the middle of his head, 'We won't hurt you, we won't hurt you.'"

Sam: "They have less to fear from children. As an adult you've learned hatred, fear, violence. They fear an adult."

Whitley: "I've sensed fear."

Sally: "Me too."

Budd: "It's very common that people say that they've felt they're afraid of us."

(The conversation then turned to the issue of sexual ex-

272

perience, disappearing pregnancies, and the sexual intrusions experienced by some of the men, involving extraction of semen with a probe, or having it drawn out with a sort of vacuum device.)

Sally: "There was somebody I met—the South American guy—he mentioned a number of dreams he'd had." (An abductee from Brazil.) "In some of the dreams he saw people. He was convinced that the people were half them and half human. They had large heads, but the features were more human. And they were children. A little boy, a little girl. I remember thinking, *Well, this was a dream.* But he had a number of dreams."

Mary: "I still keep it in my mind that it might have been something psychological with this mysterious pregnancy that I had. I did not have a false pregnancy. I was tested positive, blood test, pelvic, my periods had stopped. I was pregnant. But I keep in my mind that maybe this was some kind of psychological reaction to a miscarriage. That keeps me from going crazy, wanting my child."

Budd: "This is very hard."

Mary: "The craziest thing is that I'm not alone."

I have never before encountered such a group of seemingly ordinary people under so much pressure. They were deeply troubled by the question of what their experiences really mean.

Those who have had the experience must learn to ride a sort of psychological razor, to accept and reject at the same time. True agnosticism is a very active mental state, a sort of eager unknowing. In the direction of skepticism, for the taken, lies one form of madness; in that of belief another. One must balance between the two. For scientists there is the additional and very real danger of getting sucked into the study of a false unknown. In the case of a phenomenon as complex and yet as transient as this one, that danger is greatly intensified. But a large body of observational information now exists about the flying disks, some of it generated by skilled professional observers. And there are thousands of pages of transcript from those who may have been taken. What's more, most of them seem to have endured intrusions into their brains of one sort or another. It would seem that the existing body of data and the large number of individuals available for study might yield some useful conclusions, as long as the subject is not approached with the sort of negative hypothesis that has distorted the efforts of the debunkers.

This need for balance is fundamental to more than the process of reconciling oneself with the apparent meaning of visitation. It is also fundamental to understanding the experience. For the experience does have its symbolic center in the number three and the triangular shape. The visitors often appear in threes. They project triangular lights. They have been reported to wear various types of triangular devices and emblems. People see three pyramids or three triangles in connection with them. A huge triangular-shaped object is sometimes sighted. I had triangles etched on my arm. Dr. X and his son developed triangular rashes.

I spent fifteen years involved with the Gurdjieff Foundation, primarily because so much of the thought of G. I. Gurdjieff and his disciple P. D. Ouspensky involved the triad as a primary expression of the essential structure of life, and I have always been fascinated with the significance of this figure.

The triangle was the symbol of the ancient Triune goddess, and is, of course, viewed by Christianity as the central form of the Godhead, the Trinity.

The thirteenth-century Christian mystic Meister Eckhart said of the Holy Trinity: "God laughed, and begat the Son. Together they laughed, and begat the Holy Spirit. And from the laughter of the three the universe was born."

A current theory suggests that gravity may consist of two components counterbalancing one another, the balance of which causes the third force—which is what we call gravity—to emerge.

In order to approach an initial understanding of the visitor experience, if that is possible, it might be productive to explore the inner meaning of the triangular shape.

Triad

I began this contemplation at noon at the cabin. It was a day in early spring. I was alone. I began to think of triangles, of triads, of the struggle I have had to find a finer balance within myself.

There are many ancient traditions that view man as a being with three parts: body, mind, and heart. It seems possible that the visitors view themselves as an entire species with three parts, judging from the three distinct basic forms that have been seen. (There are variants of these forms, and also more humanlike beings, but given the degree of perceptual error that must be present here, and the fact that the three basic shapes may have many permutations, there is no way to argue with any certainty that more or less than three forms actually exist.) It is not unreasonable to consider that a species with three basic forms would choose the triangle as its most basic symbol of self, as it would express both the nature of the species and a fundamental law of structure.

To begin an exploration of this law, I would like to return to Dr. X and his seemingly enigmatic experience. This man, who was at the time of his observation a prominent physician, experienced something very peculiar. But what happened to him is not without coherence. He was awakened in the night and looked out a window of his house, which had a magnificent view of the Loire Valley and the town of Arles in France. Hanging over the valley were two glowing disk-shaped objects. Electrical discharges were taking place between them. They moved closer to Dr. X, and he observed them to merge together into a single object. Later, he discovered a triangular rash around his navel and around the navel of his son. The rashes persisted for years and, despite extensive study, remained unexplained. This case was reported and researched by Mr. Aime Michel, a

Frenchman at the time eminent in the field of UFO research. The combination of the excellent witness, the strange physical trace, and the seemingly enigmatic nature of the observation caused Mr. Michel to decide that the UFO mystery as a whole was unsolvable.

The fundamental idea of the triad as a creative energy is that two opposite forces coming into balance create a third force. The rather theatrical event that Dr. X perceived was, intentionally or not, an illustration of this principle. Could it have been a communication, even a request for some sort of response?

The imprinting of triangles on myself and Dr. X may also have a similar significance.

The idea of the triad is not static. It is an expression of a series of emanations. The third force emerges when the first and second forces come into balance, and when all three are in harmony they become a fourth thing, an indivisible whole.

I do not wish to imply by this anything beyond a human context. It is possible for man to become more whole, for each of us to make our private journey back to the place of emergence, and find there the simplest and most real of truths: that we are all at heart the same, that every body contains every soul and has room for every act without reference to its quality. There is a deep, objective awareness of self and universe that is available to all of us.

We could be part of a triad that includes the visitors. They might be the aggressive force, entering us, enforcing our passivity, seeking to draw from the relationship some new creation. But the triad can never come into harmony until there is a firm ground of understanding. We need not be blindly welcoming. What is required is objectivity. We must have a care, for if they are real it can be as persuasively argued that they are aggressive as it can be that they are benevolent. They take us in the night. They introduce their instruments and thus their reality into our brains. It is, however, too easy to call them evil, just as it is too easy to say that they are saints, kindly guides from the beyond. They are a very real and immensely complex force, the provocative nature of which demands neither hate nor love, but rather respect in a context of intellectual objectivity and emotional strength.

In ancient Taoist thought the fundamental force in the universe was duality, the yin and the yang, positive and negative, thrusting and opening, seeking and waiting, male and female. This was also thought by the Aztec and many other cultures to be fundamental to everything. And the duality, when it was in harmony, formed the triad. Throughout this chapter I will draw primarily on the Aztec imagery, as a reminder of the fragility and seriousness of our situation · if we are indeed dancing a real dance with real visitors.

The triad cannot become strong unless it is preceded by a strong duality. Without the friction of bodies there can be no child, and the universe cannot proceed. There must first be two forces, equal opposites, one that pushes and one that resists. Do the visitors perceive themselves as the aggressive force, seeking to open us to their presence? If so, everything depends on our slowly growing understanding, for unless we understand we cannot be their equal. Unless the two forces are equal, the triad will not have a chance to balance itself and the relationship will not be creative.

When the opposition between two is in balance, the third force emerges. Perhaps the French call the moment of sexual climax "the little death" because it suggests the passing of the parents and the turning of the generations. Or maybe because it is like the death of self that is involved in entering pure being, climax being a moment when the self is absorbed in ecstasy.

On the night of December 26 I felt psychologically destroyed, as if my very self had died. It could be that the basis of the fear we feel for the visitors, and—it seems to me—they for us, comes from a biological, instinctive awareness that our coming together may mean the creation of a third and greater form which will supplant us as the child does his parents.

The third force is not a small thing: It is the immense progress of life, the very movement of the universe toward whatever goal it seeks. First and second forces are people struggling in a bed. Third force is at once their frantic union and the whole urgency and implication of creation. It is their mutual attraction, the friction of their bodies—and their child.

When the internal triad of mind, body, and heart becomes fixed in a state of permanent harmony, it is because the seeker has finally died to himself and all the allures of life. Out of this death the fourth state emerges. This is the ecstatic objectivity that the Western seeker cherishes, the nirvana of the Hindu, the blooming lotus of Zen.

The human being in a state of spiritual harmony is looked upon as a sort of cosmic egg out of which hatches the bird of resurrection, which is the phoenix, also characterized as an eagle, the symbol of the yin.

In the old imagery of the tarot, and in the gospel story of the Marriage Feast at Cana, the feminine is viewed as a cup, the masculine as what is poured into it. The Aztec poets sang of the creative impact of the God and Goddess of Duality, and called the third force the song of the flower.

Distantly as I sit here on the porch at the cabin, I hear the roaring of one of our brooks, swollen by spring melt. The leaves shimmer on the trees, a swarm of Mayflies hovers in the sun. Suddenly two phoebes battle, a scream, a fluff of feathers, and then silence, both birds gone. At some moment, for a reason that seems to me to be as large and enigmatic as the universe itself, the female ceased to resist the male. The duality joined, the triad emerged, new life is now vibrant in her belly.

And both birds are a little older.

Each independent act of creation vibrates the whole web of the world, when the phoebe mates, when a woman takes the pleasure of a man.

It is hard for me to think that the relationship between two intelligent species would not be dense with creative potential.

When we were first married, Anne and I found a motto for ourselves in the Bible, from Ecclesiastes, "Two are better than one . . . and a cord of three strands is not quickly snapped." Anne cross-stitched it and it has been with us ever since. The third strand is the love, then the child, then the long unraveling of the years. At last it is what is left of a lifetime spent together, the fading remembrance, the generations to come, and also the joy that ripens in souls.

Among the Aztecs, the Lord and Lady of Duality created out of their harmony a truly extraordinary third force:

> *Man was born,*
> *sent here by our mother, our father*
> *the Lord and Lady of Duality.*

The Aztec philosophers asked the question, what is the third force, what is the bond and outcome of harmony? But they did not ask it in these structure-laden words, rather they asked, what is his flower, what is her song? The love and the child both make the marriage true.

And when the flower and the song came together, the Aztecs would say:

> *The flowers sprout, they are fresh, they grow;*
> *they open their blossoms,*
> *and from within emerge the flowers of song;*
> *among men You scatter them, You send them.*
> *You are the singer!*

The flower is the man, the song the woman, the flower of song the third strand winding gaily in the dark.

This must be a very careful business, this communion, for it can easily make such a flame that the flower is burned.

> *My flowers shall not cease to live;*
> *My songs shall never end:*
> *I, a singer, intone them:*
> *They become scattered, they are spread about.*

Cortez emerged from the sea, and the shadow of the creator as destroyer stalked in the land, then the flowers were trodden down and the brutal, beautiful Aztec civilization was destroyed forever.

Destined is my heart to vanish,
like the ever-withering flowers?
What can my heart do?
At least flowers, at least songs!

So a dark triad was completed, the gory Aztec flower cut by the booming Spanish song.

For there to be growth instead of death, much more must be brought to the triad than mere conquest. Or "contact," which, if the visitors are as advanced as they seem, would amount to a form of conquest. Communion is as wide as all the knowledge of both partners, as deep as their whole souls. Marriage requires patience, giving without thought to keep accounts. When one says "I gave this and so I am owed that," the marriage has not yet begun. Real sharing rests in a balanced recognition of sameness and difference. It is a discovery of balances and equalities.

We need to give ourselves to our experience, without knowing what it is, trusting that our understanding will grow as we proceed. To participate truly in this experience, we must marry the unknown. The only belief is the question itself: Love is a matter of leaping out into the sky.

But then again, one cannot be objective in the context of an excess of passion. We must be careful, for the stakes are high: Mankind is in the position of maturing as a species at the same time that our planet could be dying. We have a difficult road ahead. We must resist all temptation to wait for the visitors to save us. If we wait, we can be sure that nothing will save us. We have to learn to live on the edge of the razor.

When two in balance cause a third to emerge and remain in balance, something more happens: The three together become a greater whole. All seeking toward higher consciousness is a search for the sort of balance that will cause the triad to cease to be a collection of parts and become a solid.

Hidden in the Sphinx, one of our most ancient objects, is a great idea, simple and extremely powerful. To understand the riddle of the Sphinx is to know how to begin one's walk along the ancient way, the "pathless path" of old.

The most powerful moment I experienced in my search

through the modern literature about the visitors took place when I was reading *The Andreasson Affair*. Few of the accounts I have read contain as much symbolism. But this one contained a great deal, and it was quite remarkable. What interested me most about it was that Mrs. Andreasson obviously had no idea at all what it meant. But it has great meaning, and is entirely coherent in context with all I have been discussing here.

"I'm standing before a large bird," Mrs. Andreasson reported. "It's very warm. . . . And that bird looks like an eagle to me. And it's living! It has a white head and there is light in back of it—real, white light. . . . The light seems so bright in back of it. It's beautiful, bright light. . . . The light just keeps sending out rays. They keep on getting bigger and bigger. Oh, the heat is so strong!"

The great symbol of transformation, the fourth beast of the Sphinx, is the eagle. It is ever associated with heat, the energy of the sun that at once sheds the light of wisdom and the heat that burns the self away.

The riddle of the Sphinx: What has the strength of a bull, the courage of a lion, and the intelligence of a man? The answer is the Sphinx itself, who then takes wing like an eagle and looks down over life from outside of time, with true objectivity.

The flying Sphinx is a triad rendered in yet a fourth dimension of reality: a triangular solid, a pyramid, known in esoteric thought as the living eternal. The pyramids may or may not have been tombs; they were certainly symbols of the immortality of the pharaohs who built them.

Betty Andreasson had no idea what her vision was about. They asked her, "Do you understand?"

"No, I don't understand what this is all about, why I'm even here."

The central effort of my life has been the fulfillment of the triad, the creation of the eagle within me. And now I find the key myth of this transformational effort embedded in the innocent literature of the abductees, in a passage of immense power and force. Later in her transcript, Mrs. Andreasson reports being told like so many of us that she has been "chosen." This confused her, because she did not see the significance of what she had been shown. Since this image was so powerful, and so much at the center of her testimony, it seems logical that whatever it was that chose her did so to convey it.

My day alone at the cabin drew to an end. The dark came slipping out of the woods, and at length I went inside and had a meal. I did not turn on the lights, but rather chose to let the night in.

Sitting in my silent living room on the couch where the visitors left me on the night of December 26, reading through the late night hours, I reflected on the relationship between the innocent and the sublime, the new and the ancient. How is it that Mrs. Andreasson—a middle-aged American housewife with probably no access to the texts that concern themselves with this deep secret—hit upon the very symbol of the completed triad?

What old beast is shuffling toward the surface of human experience—surely not the very eagle, the Phoenix of transformation whose shadow has made me sweat with longing?

Let us turn now toward the rigorous objectivity that might allow a human spirit to take wing, to soar beyond the attractions of life that the Hindu sages have given the wonderful name *maya,* which P. D. Ouspensky in his more utilitarian interpretation described simply as "identification" with the illusionary importance of everyday affairs. It seems important to cleave to the things of life, the details of every day, but it is not very important. With care, our obsession with these things can be put aside even while our responsibility for them remains.

We do not even know if there are visitors. We do not know what we are, or why this is happening, or exactly *what* is happening. The real center of the experience lies not in some facile explanation, it lies in opening oneself to the question as it really exists, with all its mystery and danger.

To get that particular flower to open a little more, it may be helpful to turn to another enigma: the tarot. First, please set aside any notion of fortune-telling. I suggest that the Major Arcana reveal a hidden symbolic coherence of great purity that has more to do with order than chance. About fifteen years ago I became interested in the tarot when I was studying the rise of monasticism in Europe for a historical novel I never actually wrote. I came to realize that the tarot is much more than a deck of fortune-telling cards; it is a sort philosophical machine that presents its ideas in the form of pictures rather than words.

The story it tells is an interesting one: The face cards of the tarot—the Major Arcana—can be arranged in such a manner that they work as signposts toward spiritual evolution. Card twenty-one, the final card in this arrangement, is called the World. There is in this card as deep a representation of the spirit and force of the triad as exists on earth: It consists of the four beasts of the Sphinx, one at each corner, surrounding an enigmatic and powerful figure who stands within a wreath.

This figure's genitals are hidden by a cloth; it has breasts and also the suggestion of a male form. It is intended, I think, to represent a potential in all human beings. In the hands are carried tools from the table of the card of the Magician, most specifically, his wand. The figure may represent humankind transformed, the beast reborn, man or woman, half human and half god.

Out of communion there emerges transformation. The strong body, the brave heart, and the intelligence of a human being.

What is the true aim of mind? Does it seek knowledge only for the sake of constructing the intricate pleasures and poisons of technology, or just to know, or is there another dancer in the shadows?

The mind can bring understanding to body and heart. It can direct the body toward healthy ways and the heart toward the balm of insight.

Then, when a being comes into harmony, body, mind, and heart reconciled, there is a chance to look up from the plodding and the toil, and there to encounter the whole tremendous ecstasy that made the phoebes scream in the morning light.

It is then that the struggling worm turns from clay into fire and the soul takes wing, sweeping out of time and chance, going higher and higher, daring like the eagle the relentless, revealing fire of the sun.

In all of us there is an unaddressed urgency which we cannot really name, which seems to lie at the heart of our hope for ourselves. It is the flight of the eagle that we seek, the walk of the monk down the nirvana road, the faith of the old priest whose mass in his humble little church last Sunday was really said between the worlds of body and soul.

We go from the endless battle of duality to the harmony of the triad, and then to the mystery of the eagle. Each of us is potentially a transfigured being, the friend of God, the Phoenix gliding free.

Are the visitors asking us to form a triad with them? Is that the purpose of all the triangular and pyramidal imagery? Maybe, but there may be another truth here. Maybe I have located these symbols in the visitor material simply because they are so central to my own life. My whole soul and breath are devoted to the groping toward transformation. Maybe it is inevitable that I would extract relevant material from any enigma that I encounter.

And yet, I cannot think that Betty Andreasson chose to present the fiery drama of the risen eagle with such force in her testimony entirely on her own.

Life is expressed in Christian cosmology as having emerged from the unity of the Trinity. And this is nothing mystical, it is very simply true: We could not exist without all three of the dimensions that sustain us. Solids depend on breadth, depth, and height, and we cannot construct a perceptual reality sufficiently reflective of all potential with fewer dimensions. The advantage of three dimensions is that they allow parallel movement through space and time, which is essential to experience.

Everybody is afraid of giving up themselves and just being. The fire behind the eagle felt so hot that it terrified Betty Andreasson, for it was the fire that consumes the self. It was the apocalypse of the soul.

Our agony is to stand before the utter blackness of the unknown with full knowledge that there is something there, and it is alive, and if one is to remain on the path of inner search, one must trust it even though it may very certainly be dangerous. Strength is needed to endure the fire, courage to enter it, intelligence to come out alive.

There has ever been in the life of man this idea of the triad as the primary force of growth. The Sphinx is a very old construction, and the sacred graphic known as the Kali yantra or Primordial Image in the Indian Tantras may be older still. This ancient symbol, a triangle with the *bindu*, or spark of life, at the center, is associated with the Triple Goddess who rules past, present, and future (length, breadth, and depth), and the trimesters of pregnancy, and the three seasons of life: childhood, maturity, and age.

Out in my woods the hemlocks are sighing with the long breath of the night.

Three Ways was one name of the Roman goddess Hecate, whose three-faced image at crossroads traditionally received offerings of cake, fruit, and money. Money is still offered at one of her ancient shrines, the Trevi Fountain, and the traditional idea of being blessed by throwing three coins in the fountain has persisted into our era.

The Irish god Trefuilngid is the patron of the trefoil, or three leaf, the shamrock. Trefuilngid is known as the Triple Bearer of the Triple Key, which is the same appellation given Shiva, Astarte, and Ishtar, three ancient manifestations of the Triune Goddess. Of course, the shamrock is also the symbol of St. Patrick. Among the ancient Arabs the trefoil was called the *shamrakh*, which was a symbol of the male trident of fertility. Did the Irish once know the Arabs? What dark seas flowed before we learned how to write our history down, and what triumphs have been swallowed by their waters?

In the Greek alphabet the fourth letter, delta, is the symbol of the Holy Door, and among the Egyptians the triangle was the symbol of Men-Nefer, the very ancient goddess of the mother city Memphis, as identified in the Egyptian Book of the Dead. The object of the worship of the Yantra is to attain unity with the Mother of the Universe in Her forms as Mind, Life, and Matter . . . preparator to Yoga union with Her as She is in herself as Pure Consciousness.

To the Gnostics of the later Roman Empire, the triangle

signified creative intellect, the balancing coolness of mind, the persistent reconciliation that proceeds in the souls of the ones who seek Christ within.

Among the ancient Sumerians Ishtar was the Triune Goddess, as among modern Christians the Trinity is the central figure of the creative force, God the Father, God the Son, and God the Holy Spirit.

The Way to Christ is lit by the Star of Bethlehem, as also the word *Ishtar* means "star." Babylonian scriptures call Ishtar the Light of the World, Opener of the Womb, Leader of Armies, Lawgiver, Forgiver of Sins. It is from the legend of Ishtar that the story of the descent into the underworld enters so many of man's traditions: The giver of life overcomes the taker of life, and rebirth emerges.

Shadrach, Meshach, and Abendigo traveled the path of transformation, and were not consumed in its fires. They symbolically sat astride the shoulders of the eagle, as did the medicine men of Native America, the shamans of the Siberian plains, the witches of old Europe when the ice king strode half the continent.

Transformation: Among the Apache, when someone was called in his heart to be a shaman, it meant that he had to go into the world of the dead, he had to dare the fire. The Apache would find a cliff and jump off. If he lived, he became a man of medicine. If he died, he died.

Length and breadth, when they merge, make the solid, which the Pythagoreans thought of as mind emerging into reality. "Know thyself," was ever the refrain of the old Greek mystical philosophers, the admonition of Apollo and of the visitors who took Betty Andreasson.

Could it be that we are about to rediscover the actual, physical truth of the ancient idea that to know the mind is to know the universe? A child reported: "The unconscious mind is like the universe beyond the quasars. It's a place we want to go to find out what's there."

What is there? Could there be in the tinkling caves of the soul a door that leads beyond the edge of reality? Is that where the eagle will go when it takes flight? Is that where the visitors came from?

No matter what factual reality attaches to the experience of the visitors, the presence of the triangle as a frequently observed symbol by many people who have no idea at all of its rich heritage suggests that there is much here that is worth exploring, and not only with the tools of science but also with the tools of myth and philosophy, with the heart as well as the intellect.

At the least, we are dealing with a grand flight of our own mind, in some mysterious way collecting itself against thunder from the future, seeking in this time of world danger toward its oldest truth.

It could be that the triangle performs a similar symbolic function wherever the three-dimensional universe prevails. In speaking to us through that particular form—if that is what they are doing—the visitors may be announcing themselves as belonging to the same laws that inscribe our beings, and identifying their presence with the everlasting ideology of transformation. It could be a symbol not only of mutuality of structure but of shared aim, which is the continuation of life and the search for wisdom. The additional presence of the eagle in this symbology confirms its sense and direction and makes it difficult to assert that it is all just an accident, as anchored in chance as a remarkable thing I once saw during a terrible storm at sea.

I was in a small cabin cruiser and a squall had come up out of nowhere on the Gulf of Mexico. The wind was gusting terrifically and twenty-foot waves were threatening to swamp the boat. Should the engine get flooded by water coming in over the transom, we had very little chance of survival.

Slowly, painfully, the boat climbed the wall of a wave. As we teetered over, I saw across the sweep of whitecaps an amazing sight: In the middle of a shaft of bright sunlight stood a tall pyramidal wave. Its sides were glassy smooth, its peak streaming white spray into the wind. For a moment it seemed perfect, solid, and eternal. Then it sank slowly down, the victim of the same random forces that had created it. As we limped back into Port Aransas after the storm the captain commented, "You see some strange things out there."

EPILOGUE

Where can such a journey as this end? Swaying in the wind with the stars, or along the dark strand of faith? For me it must be in the human dimension, for that is the only place where we can hope to make progress toward a fuller understanding.

I do not have it in me to be a believer. Nor can I be a true skeptic, for I loathe the narrow and love the broad. I cannot say, in all truth, that I am certain the visitors are present as entities entirely independent of their observers. Nor can I say that I do not think they are here at all.

It is not enough for us to ascribe the visitor experience to some unusual manifestation of known phenomena and then ignore it. Science has not explained the visitors; indeed, it has not even begun to explore them. Nor is it sufficient to say that "higher" beings are studying us, and remain passively waiting for what tidbits of knowledge they may toss us.

A striking fact of the colloquy was the general expectation of catastrophe, a possibility that I also fear. Throughout the literature of abduction, there is a frequent message of apocalypse. This is also the message of Fundamentalism— and of some parts of both the peace and environmental movements.

Presently in the Soviet Union there is so much fear regarding the "end of days" that Premier Gorbachev has taken to wearing makeup on Soviet television to obscure a birthmark on his forehead because people have called it "the mark of the beast." And there has been an undertone of fear in relation to the fact that the word *chernobyl* in Ukranian means "wormwood," the name of the star in the Book of Revelation that is poison to a third of the waters.

As the ninth century closed toward the first millennium, the Western world was also swept by wave after wave of millennial hysteria. The end of days was expected momentarily. Of course, they did not have nuclear weapons and an expanding hole in the ozone layer.

The Celtic tradition held that the veil between the worlds of man and spirit could grow thin at certain seasons of the year. Maybe there are seasons of ages, and the veil between matter and mind is now growing thin. It could be that thought is beginning to cross into the concrete world, or even that mind is learning how to manipulate reality to its own secret ends.

Why did an actor report that a UFO was the source of the great New York City blackout, considering that the first "UFO-induced" blackout took place in a play? And the first recorded instance of a UFO killing a car's engine seems to have been in a work of fiction.

Understand, I am not dismissing the phenomenon with yet another version of the "wind making the stars sway" explanation. No, the visitors may very well be real. Quite real. But what are they, and what in their context does the word *real* actually mean? I do not think that this is a question that will in the end admit itself to a linear and mechanistic answer.

The visitor experience drives us to extremes. Those who have seen the devices or their occupants are often convinced that they are extraterrestrial in origin. And science debunks that, just as the clergy debunked the fairies in centuries past, and for the same reason: These outrageous enigmatics so threaten the established order of belief that they must at all costs be rejected. There was no room in seventeenth-century Christian theology for fairies, and there is scarcely more in twentieth-century scientific theory for visitors as peculiar as these.

But the difference between science and theology is that science can make room for new things. In 1932 Dr. Albert Einstein stated, "There is not the slightest indication that nuclear energy will ever be obtainable. It would mean that the atom would have to be shattered at will." Within the decade, he would write President Roosevelt the letter that started the United States on the road to splitting the atom and creating the atomic bomb.

"I know in my heart that they might possibly be here," one skeptic said to me, "but if they are, and they're acting like this, it's just so damned disappointing that I prefer not to believe it."

Why disappointing? Why not leave the question open for now? True, the visitors have not come down from their spaceships and given us a holographic atlas of their "home planet" . . . or the secret of their interstellar drive. They are very strange. But that could be a two-way street. We may also be very strange.

Looking back over my experiences with the visitors, I cannot say that I felt inferior to them. On the contrary, the people I encountered did not seem superior so much as wiser, but also more simple and unformed as individuals. And they not only feared me, they seemed in awe of me. When they asked me what they could do to help me stop screaming, there was an edge of panic among them.

They are not all-powerful superbeings. They are frail, limited individuals far from home—if indeed this world is not their home.

I can discern a visible agenda of contact in what is happening. Over the past forty or so years their involvement with us has not only been deepening, it has been spreading rapidly through the society. At least, this is how things appear. The truth may be that it is not their involvement that is increasing, but our perception that is becoming sharper.

The evolution of this increase in perception may have a very definite design. We initially noticed objects from afar, then closer, then we remembered seeing the visitors, and now we are beginning to remember being with them. More than one of those taken were told that he or she would remember nothing "for five years," or "until 1984," or "in a few years." Will the visitors emerge into our world on a flood of memory?

And if so, then why? Why not simply land, open the hatch and come out? It could be that they wish to avoid what Cortez did with such eagerness. It is not difficult to crush the flower of a culture's spontaneity. A friend of mine sat in a Native American medicine circle within hearing distance of a hissing interstate highway—and he heard the emptiness in the old chants, the sadness where conviction once rang. And no new stories are being woven in Papua, New Guinea. The streets are becoming a ramshackle version of Lansing, Michigan. It's all turning into rock 'n' roll, the scepters of the kings are being broken up for firewood and the old, rich truths of that culture now seem to its inheritors an embarrassment.

Would we not all risk being lost in nonmeaning if an apparently superior visitor culture emerged suddenly into the open? Science, religion, even the arts might be shattered by the appearance of a culture that already knew everything we want to know about the universe.

Unless, of course, it were to emerge not into blinded awe, but into our understanding of its truth, its strengths as well as its weaknesses.

Maybe that is why two triangles were inscribed on my arm: to symbolize that we are each a tiny, complete universe, a small but valid version of the whole; that the smaller is not less perfect than the greater, but only less mature.

Beyond the present level, what awaits? What will happen in ten years, in twenty years? As we begin to admit that the visitor phenomenon has some sort of extrapersonal reality, perhaps it will begin to come into clearer focus.

A visitor once said to an abductee, "On is off and off is on. We confuse the language." There is something of the mirror image in all this, and in the visitors more than a little of the prankster which has so much significance in our own mystical literature. From Till Eulenspiegel to Mullah Nasir Eddin, we have always accepted pranks as one output of true understanding. "God laughs and plays," said Meister Eckhart.

Dr. David M. Jacobs of Temple University recounts a fascinating story of a woman in Philadelphia who once saw a flying disk across the street from her house. As soon as she looked at it, it came closer. She beheld a row of nine windows. In one window stood a man with a cigar in his mouth, staring out. He was as motionless as a statue. The next window revealed a woman in a flowered dress sitting in a chair, also in some sort of trance. Three of the small gray beings then came past the other windows and conducted the entranced woman from the chair and off down a corridor.

The woman who saw this was not a "saucer nut," not a psychotic, but rather an ordinary person. She neither sought nor received fame or money. She simply told what she saw, all innocent to the fascinating combination of absurdity and far-reaching implication in her story. At the moment of highest absurdity and intensity in my experience, when I was called their chosen one, I had a distinct memory of seeing them getting a woman in a flowered dress very excited with a similar speech.

How can something so profound and even dangerous as the visitor experience also be so ludicrous? It would seem to me to be possible to say that the mind, also, laughs and plays.

In August 1986 a man had a remarkable experience while driving toward Great Neck, Long Island, on the Grand Central Parkway. It was 9:30 P.M., and the sky was overcast, with a three-thousand-foot ceiling. The man suddenly saw an enormous airplane coming toward him so low that it looked like it was about to land on the parkway. It had two bright lights in the nose, lights that seemed to shine beams directly into his eyes. There was a red light at the tip of one wing and a green one at the tip of the other. As he passed under the plane he looked up and saw that there were rivets in the undersurface, which was streaked as if it had scraped along the ground. He saw four engine pods with whirling propellers. The nose of the plane was flat and there was no horizontal stabilizer. The man slowed down and leaned his head out the window, looking up at the bizarre craft. It seemed almost to be standing still, and the propellers made no noise at all. Soon he was past it and taking the Great Neck exit, which makes a horseshoe around a small hill. Above this hill he saw what appeared to be an advertising sign made of many small lightbulbs. It had an angle in it, suggesting that it was attached to two sides of a building. It was flashing, but the symbols were incomprehensible. As he rounded the horseshoe and saw the sign from another angle, he realized that there was no building there. Then he concluded that it must be a plane. But it suddenly shot off to the southwest, rising into the overcast with blinding speed.

What happened to this man? What did he see? It would be easy to dismiss his experiences as a pair of hallucinations. Easy to debunk this one.

There is, however, a problem. The problem is the man himself, and his extraordinarily apt qualifications. He is a leading perceptual psychologist with encyclopedic knowledge of just exactly how the brain perceives things, and what misperceptions mean. What's more, he has a near-photographic memory and eyes so superb that he can see the moons of Jupiter unaided. He is also highly intelligent and exceptionally emotionally stable, having had many hundreds of hours of psychoanalysis as part of his clinical training.

Anybody can have a hallucination or a misperception. But this highly qualified man feels certain that what he saw was actually there. Interestingly, other drivers did not react to it at all. I wonder if that might not be because they *believed* an illusion that this man's mind was too highly trained to mistake. Most people saw a plane and an advertising sign. But this acute, trained mind saw beyond the camouflage to what was really there—a device of unknown origin and purpose.

How interesting that such an outrageous perceptual joke would be played on a skilled perceptual scientist—who himself has superb perceptual equipment. Or perhaps the visitors were indifferent, and chance played the joke. Then again, maybe it wasn't a joke at all. What about the light shining in his eyes—did they use it to learn something from him, or induce him to act in some drama of importance to their enigmatic designs?

I would not be at all surprised if the visitors are real and are slowly coming into contact with us according to an agenda of their own devising, which proceeds as human understanding increases. If they are not from our universe it could be necessary for us to understand them *before* they can emerge into our reality. In our universe, their reality may depend on our belief. Thus the corridor into our world could in a very true sense be through our own minds.

The idea of parallel universes is neither proven nor new. It has a distinguished history in physics. Of course, the conditions under which movement between universes might be possible are not known.

We have seen that the visitors are not fairies, and that their ships are not figments of the wind. Now that we know this, what more will we learn?

Humanity could be clutching the frail barque of an outmoded world view while the wind of the mind is swaying the stars into very real craft, and out of them is coming . . . a faint call for help from a lady in a flowered dress.

This is not a "mere" matter, to be explained away by one facile dodge or another. It is an immense human reality, vast in its impact and complexity. It has coherence, strange but undeniable, and thus there is certainly a process of thought that will draw it into our understanding. Presently it is lacking effective definition. To leave it this way when it seems so rich with potential would be a shame. But has science the wit to study such an elusive and multidimensional enigma?

I say yes. Resoundingly so. Even a brilliant but arrogant curmudgeon of my acquaintance, who denounced it all as "preposterous," is important to an understanding of it. Blind denial is as empty a response as blind acceptance, and operates on the same level of validity. There is no real intellectual difference between the haughty psychiatrist or physicist and his refusal to accept the truth, and the nervous "contactee" eager to see the phenomenon as a dimensionless cartoon of space friends. We must break through both distortions—and we certainly can.

The visitor experience may be our first true quantum discovery in the large-scale world: The very act of observing it may be creating it as a concrete actuality, with sense, definition, and a consciousness of its own. And perhaps, in their world, the visitors are working as hard to create us.

Truly, such an act of mutual insight and courage would be communion . . . two universes spinning each other together . . . the old weaver of reality rethreading creation's loom. Who knows, maybe really skilled observation and genuine insight will cause the visitors to come bursting to the surface shaking like coelacanths in a net.

Something is here, be it a message from the stars or from the booming labyrinth of the mind . . . or from both. It must have left a signature somewhere, a thread in the snow, the scratch of a strange nail upon a wall. And we can certainly find that thread, if we bring humor and honesty and courage—and great care—to the effort. In taking the thread we might find ourselves in possession of a very real key to the universe. In any case, accepting that the visitor experience is not a false unknown will relieve a lot of suffering.

Once the thread is in hand, our own mythology will tell us where it leads, for it will be the same thread that the maiden Ariadne handed to Theseus when he stood before the maze of the Minotaur, young and strong and mad with courage.

And we will all go down the labyrinth, to meet whatever awaits us there.

APPENDIX I
A Statement from Donald F. Klein, MD

I have examined Whitley Strieber and found that he is not suffering from a psychosis. He is not hallucinating in a manner characteristic of psychosis. I also see no evidence of an anxiety state, mood disorder, or personality disorder. He is an excellent hypnotic subject, who appeared to make an honest attempt while under hypnosis to describe what he remembered. He has approached the dilemma of what is happening to him in a careful and forthright way and has pursued his investigation with diligence. After an initial period of stress, he became much more calm about his situation and soon learned to deal with it in a psychologically healthy way. He appears to me to have adapted very well to life at a high level of uncertainty.

—DONALD F. KLEIN, MD
Director of Research
New York State Psychiatric Institute

APPENDIX 2
Polygraph Results

On October 31, 1986, I was polygraphed by Ned Laurendi, president of the Society of Professional Investigators and vice-president of the Empire State Polygraph Society. He has been a polygraph operator for twenty-five years. He was paid in the normal course of business before the polygraph was undertaken, and was not familiar with me before the test.

I was aware of the controversial nature of polygraphic results, and so determined to add a test of Mr. Laurendi's effectiveness to the process. Without telling him, I lied in my answers to control questions thirteen and sixteen. In both cases, he detected the lie correctly. His ability to do this—along with his leading position in his field—has convinced me that, despite the controversies surrounding lie detection in general, he is very skilled indeed.

The letter referred to in the test results was written to Mr. Laurendi on October 17, 1986, and outlined the part of my experiences that I remembererd before hypnosis.

The reason I carried out the polygraph was to reassure readers that I honestly think that I perceived the things reported in this book. It is not fiction, and does not contain a word of fiction. My successful completion of this test in no way proves that my recollection of my experiences is correct, but it does confirm that I have described what I saw to the best of my ability.

Test Results
1. Are you known as Whitley Strieber?
   Yes. (Evaluated true.)

2. Do you intend to answer truthfully?
   Yes. (Evaluated true.)
3. Did you intentionally plan to be given a lie detector test on this Halloween day? (Note: Mr. Laurendi suspected that he might be the victim of a practical joke, an understandable response in view of the coincidence of date and the nature of my experience.)
   No. (Evaluated true.)
4. Do you belong to any cults?
   No. (Evaluated true.)
5. Do you think that those things happened to you on October 4, 1985, that were outlined in your letter dated October 17, 1986?
   Yes. (Evaluated true.)
6. Did you ever fraudulently conceal any information prior to 1984?
   No. (Evaluated true.)
7. Do you think those things happened to you on December 26, 1985, as outlined in your letter dated October 17, 1986?
   Yes. (Evaluated true.)
8. Besides when you were eight years old, did you ever hallucinate again prior to 1984? (Note: I had a hallucination during a fever when I was eight.)
   No. (Evaluated true.)
9. Do you live in New York?
   Yes. (Evaluated true.)
10. Do you think those things happened to you on March 15, 1986, as outlined in your letter dated October 17, 1986?
    Yes. (Evaluated true.)
11. Did you ever lie for personal gain prior to 1984?
    No. (Evaluated true.)
12. Do you think those things happened to you in late March 1986 as outlined in your letter dated October 17, 1986? (This referred to the intrusion of the needle into my nose and my visit to the doctor to have the injury examined.)
    Yes. (Evaluated true.)

13. Did you ever lie to anyone in a business venture prior to 1984?
    No. (Evaluated as possibly untrue. A correct response. I've been in business for twenty years and I'm not sure I haven't lied occasionally.)
14. Have you lied to those people who interviewed you with respect to the four items we discussed?
    No. (Evaluated true.)
15. Are you a free-lance writer?
    Yes. (Evaluated true.)
16. Did you ever lie to someone that trusted you prior to 1984?
    No. (Evaluated false. Again, he was right. I lied to my parents, for example, when they asked me if I knew what could have burned the house down when I was a boy.)
17. Have you ever intentionally ingested any hallucinogenic drug?
    No. (Evaluated true.)
18. Did you ever take any prescription medication without a doctor's permission?
    No. (Evaluated true.)

Note: Some control questions are dated "prior to 1984" to more clearly separate my responses to them from surrounding key questions. They are not meant to imply that my responses would have been any different after 1984.

Readers should address correspondence as follows:

Whitley Strieber
Box 188
496 La Guardia Place
New York, New York 10012